Student's Solutions Manual for
Physical Chemistry

Thomas Engel
University of Washington

Philip Reid
University of Washington

San Francisco Boston New York
Cape Town Hong Kong London Madrid Mexico City
Montreal Munich Paris Singapore Sydney Tokyo Toronto

ISBN 0-8053-3848-9

1 2 3 4 5 6 7 8 9 10–B&B-09 08 07 06
www.aw-bc.com

Preface

This *Student's Solutions Manual* is intended to provide you with examples of good problem-solving techniques and strategies. To achieve that end, the solutions presented extend the range of worked-out problems beyond the Example Problems in the text. We recommend you try to solve each problem on your own before you read the solution. Simply reading solutions, without first struggling with the issues, has limited educational value.

As you work through each solution, make sure you understand how and why each step is taken. See if you can understand which aspects of the problem made this solution strategy appropriate. You will be successful on exams not by memorizing solutions to particular problems but by coming to recognize which kinds of problem-solving strategies go with which types of problems.

We have made every effort to be accurate and correct in these solutions. However, if you do find errors or ambiguities, we would be very grateful to hear from you. Please contact us at: pchem@chem.washington.edu.

Thomas Engel
University of Washington
Philip Reid
University of Washington

Chapter 1: Fundamental Concepts of Thermodynamics

P1.1) A sealed flask with a capacity of 1.00 dm^3 contains 5.00 g of ethane. The flask is so weak that it will burst if the pressure exceeds 1.00×10^6 Pa. At what temperature will the pressure of the gas exceed the bursting temperature?

$$T = \frac{PV}{nR} = \frac{1.00\times10^{6}\,\text{Pa}\times10^{-3}\,\text{m}^{3}}{\dfrac{5.00\times\text{g}}{30.07\text{ g mol}^{-1}}\times8.314\text{J mol}^{-1}\text{K}^{-1}} = 723\text{ K}$$

P1.4) A compressed cylinder of gas contains 1.50×10^3 g of N_2 gas at a pressure of 2.00×10^7 Pa and a temperature of 17.1°C. What volume of gas has been released into the atmosphere if the final pressure in the cylinder is 1.80×10^5 Pa? Assume ideal behavior and that the gas temperature is unchanged.

Let n_i and n_f be the initial and final number of mols of N_2 in the cylinder.

$$\frac{n_iRT}{P_i} = \frac{n_fRT}{P_f}$$

$$n_f = n_i\frac{P_f}{P_i} = \frac{1.50\times10^{3}\text{g}}{28.01\text{g mol}^{-1}}\times\frac{1.80\times10^{5}\text{Pa}}{2.00\times10^{7}\text{Pa}} = 0.482\text{ mol}$$

$$n_i = \frac{1.50\times10^{3}\text{g}}{28.01\text{g mol}^{-1}} = 53.55\text{ mol}$$

The volume of gas released into the atmosphere is given by

$$V = \frac{(n_f - n_i)RT}{P} = \frac{(53.55-0.482)\,\text{mol}\times8.2057\times10^{-2}\text{L atm mol}^{-1}\text{K}^{-1}\times290.2\text{ K}}{1\,\text{atm}}$$

$$= 1.26\times10^{3}\text{ L}$$

P1.7) A mixture of 2.50×10^{-3} g of O_2, 3.51×10^{-3} mol of N_2, and 4.67×10^{20} molecules of CO are placed into a vessel of volume 3.50 L at 5.20°C.

a) Calculate the total pressure in the vessel.

b) Calculate the mole fractions and partial pressures of each gas.

a)

$$n_{O_2} = \frac{2.50\times10^{-3}\text{g}}{32.0\text{ g mol}^{-1}} = 7.81\times10^{-5}\text{mol};\; n_{CO} = \frac{4.67\times10^{20}\text{molecules}}{6.022\times10^{23}\text{molecules mol}^{-1}} = 7.75\times10^{-4}\text{mol}$$

$$n_{total} = n_{O_2} + n_{N_2} + n_{CO} = 7.81\times10^{-5}\text{mol} + 3.51\times10^{-3}\text{mol} + 7.75\times10^{-4}\text{mol} = 4.36\times10^{-3}\text{mol}$$

$$P_{total} = \frac{nRT}{V} = \frac{4.36\times10^{-3}\text{mol}\times8.314\times10^{-2}\ \text{L bar mol}^{-1}\text{K}^{-1}\times278.3\ \text{K}}{3.50\ \text{L}} = 2.88\times10^{-2}\text{bar}$$

b)

$$x_{O_2} = \frac{7.81\times10^{-5}\text{mol}}{4.36\times10^{-3}\text{mol}} = 0.0179;\ x_{N_2} = \frac{3.51\times10^{-3}\text{mol}}{4.36\times10^{-3}\text{mol}} = 0.803;$$

$$x_{CO} = \frac{7.75\times10^{-4}\text{mol}}{4.36\times10^{-3}\text{mol}} = 0.178$$

$$P_{O_2} = x_{O_2}P_{total} = 0.0179\times2.88\times10^{-2}\text{bar} = 5.16\times10^{-4}\text{bar}$$

$$P_{N_2} = x_{N_2}P_{total} = 0.803\times2.88\times10^{-2}\text{bar} = 2.31\times10^{-2}\text{bar}$$

$$P_{CO} = x_{CO}P_{total} = 0.177\times2.88\times10^{-2}\text{bar} = 5.10\times10^{-3}\text{bar}$$

P1.9) A rigid vessel of volume 0.500 m^3 containing H_2 at 20.5°C and a pressure of 611× 10^3 Pa is connected to a second rigid vessel of volume 0.750 m^3 containing Ar at 31.2°C at a pressure of 433× 10^3 Pa. A valve separating the two vessels is opened and both are cooled to a temperature of 14.5°C. What is the final pressure in the vessels?

$$n_{H_2} = \frac{PV}{RT} = \frac{611\times10^3\text{Pa}\times0.500\text{m}^3}{8.314\ \text{J mol}^{-1}\text{K}^{-1}\times(273.15+20.5)\ \text{K}} = 125\ \text{mol}$$

$$n_{Ar} = \frac{PV}{RT} = \frac{433\times10^3\text{Pa}\times0.750\text{m}^3}{8.314\text{J mol}^{-1}\text{K}^{-1}\times(273.15+31.2)\ \text{K}} = 128\ \text{mol}$$

$$P = \frac{nRT}{V} = \frac{(125+128)\ \text{mol}\times8.314\text{J mol}^{-1}\text{K}^{-1}\times(273.15+14.5)\ \text{K}}{(0.500+0.750)\ \text{m}^3} = 4.84\times10^5\text{Pa}$$

P1.11) A glass bulb of volume 0.136 L contains 0.7031 g of gas at 759.0 Torr and 99.5°C. What is the molar mass of the gas?

$$n = \frac{m}{M} = \frac{PV}{RT};\ M = m\frac{RT}{PV}$$

$$M = 0.7031\text{g}\times\frac{8.2057\times10^{-2}\text{L atm mol}^{-1}\text{K}^{-1}\times(273.15+99.5)\ \text{K}}{\frac{759}{760}\text{atm}\times0.136\ \text{L}} = 158\ \text{amu}$$

P1.13) A gas sample is known to be a mixture of ethane and butane. A bulb of 200.0-cm^3 capacity is filled with the gas to a pressure of 100.0 × 10^3 Pa at 20.0°C. If the weight of the gas in the bulb is 0.3846 g, what is the mole percent of butane in the mixture?

$n_1 =$ moles of ethane $\quad n_2 =$ moles of butane

$$n_1 + n_2 = \frac{PV}{RT} = \frac{100.0\times10^3\,\text{Pa}\times0.200\times10^{-3}\text{m}^3}{8.314\text{ J mol}^{-1}\text{ K}^{-1}\times293\text{ K}} = 8.21\times10^{-3}\text{mol}$$

The total mass is

$$n_1M_1 + n_2M_2 = 0.3846\text{ g}$$

Dividing this equation by $n_1 + n_2$

$$\frac{n_1M_1}{n_1+n_2} + \frac{n_2M_2}{n_1+n_2} = \frac{0.3846\text{ g}}{8.21\times10^{-3}\text{mol}} = 46.87\text{ g mol}^{-1}$$

$$x_1M_1 + x_2M_2 = (1-x_2)M_1 + x_2M_2 = 46.87\text{ g mol}^{-1}$$

$$x_2 = \frac{46.87\text{ g mol}^{-1} - M_1}{M_2 - M_1} = \frac{46.87\text{ g mol}^{-1} - 30.069\text{ g mol}^{-1}}{58.123\text{ g mol}^{-1} - 30.069\text{ g mol}^{-1}} = 0.599$$

mole % = 59.9%

P1.16) Assume that air has a mean molar mass of 28.9 g mol^{-1} and that the atmosphere has a uniform temperature of 25.0°C. Calculate the barometric pressure at Denver, for which $z = 1600$ m. Use the information contained in Problem P1.15.

$$P = P^0 e^{-\frac{M_i gz}{RT}} = 10^5\text{Pa}\exp\left(-\frac{28.9\times10^{-3}\text{kg}\times9.81\text{ m s}^{-2}\times1600\text{ m}}{8.314\text{ J mol}^{-1}\text{ K}^{-1}\times300\text{ K}}\right) = 8.34\times10^4\text{ Pa}$$

P1.18) Calculate the pressure exerted by benzene for a molar volume 1.42 L at 790 K using the Redlich-Kwong equation of state:

$$P = \frac{RT}{V_m - b} - \frac{a}{\sqrt{T}}\frac{1}{V_m(V_m+b)} = \frac{nRT}{V-nb} - \frac{n^2a}{\sqrt{T}}\frac{1}{V(V+nb)}$$. The Redlich-Kwong parameters a and b for benzene are 452.0 bar dm^6 mol^{-2} K$^{1/2}$ and 0.08271 dm^3 mol^{-1}, respectively. Is the attractive or repulsive portion of the potential dominant under these conditions?

$$P = \frac{RT}{V_m - b} - \frac{a}{\sqrt{T}}\frac{1}{V_m(V_m+b)}$$

$$= \frac{8.314\times10^{-2}\text{bar dm}^3\text{mol}^{-1}\text{K}^{-1}\times790\text{ K}}{1.42\text{ dm}^3\text{mol}^{-1} - 0.08271\text{dm}^3\text{mol}^{-1}}$$

$$-\frac{452.0\text{ bar dm}^6\text{mol}^{-2}\text{K}^{\frac{1}{2}}}{\sqrt{790\text{ K}}}\times\frac{1}{1.42\text{ dm}^3\text{mol}^{-1}\times\left(1.42\text{ dm}^3\text{mol}^{-1}+0.08271\text{dm}^3\text{mol}^{-1}\right)}$$

$$P = 41.6\text{ bar}$$

$$P_{ideal} = \frac{RT}{V} = \frac{8.3145\times10^{-2}\times\text{L bar mol}^{-1}\text{K}^{-1}\times790\text{ K}}{1.42\text{ L}} = 46.3\text{ bar}$$

Because $P < P_{ideal}$, the attractive part of the potential dominates.

Chapter 2: Heat, Work, Internal Energy, Enthalpy, and the First Law of Thermodynamics

P2.1) 3.00 moles of an ideal gas at 27.0°C expands isothermally from an initial volume of 20.0 dm^3 to a final volume of 60.0 dm^3. Calculate w for this process a) for expansion against a constant external pressure of 1.00 × 10^5 Pa and b) for a reversible expansion.

a) $w = -P_{external}\Delta V = -1.00\times 10^5\text{Pa}\times(60.0\text{-}20.0)\times 10^{-3}\text{m}^3 = -4.00\times 10^3\text{J}$

b)

$$w_{reversible} = -nRT\ln\frac{V_f}{V_i} = -3.00\text{mol}\times 8.314\text{ J mol}^{-1}\text{K}^{-1}\times 300\text{ K}\times\ln\frac{60.0\text{ dm}^3}{20.0\text{ dm}^3} = -8.22\times 10^3\text{J}$$

P2.4) For 1 mol of an ideal gas, $P_{external} = P = 200 \times 10^3$ Pa. The temperature is changed from 100°C to 25.0°C, and $C_{V,m} = 3/2R$. Calculate q, w, ΔU, and ΔH.

$$\Delta U = nC_{V,m}\Delta T = \frac{3}{2}\times 8.314\text{ J mol}^{-1}\text{K}^{-1}\times(298\text{ K} - 373\text{ K}) = -935\text{ J}$$

$$\begin{aligned}\Delta H = nC_{P,m}\Delta T &= n(C_{V,m} + R)\Delta T\\ &= \frac{5}{2}\times 8.314\text{ J mol}^{-1}\text{K}^{-1}\times(298\text{ K} - 373\text{ K})\\ &= -1.56\times 10^3\text{J}\\ &= q_P\end{aligned}$$

$$w = \Delta U - q_P = -935\text{ J} + 1.56\times 10^3\text{J} = 624\text{ J}$$

P2.6) Calculate ΔH and ΔU for the transformation of 1 mol of an ideal gas from 27.0°C and 1.00 atm to 327°C and 17.0 atm if $C_{P,m} = 20.9 + 0.042\frac{T}{\text{K}}$ in units of J K^{-1}mol^{-1}.

$$\begin{aligned}\Delta H &= n\int_{T_i}^{T_f} C_{P,m}dT\\ &= 1\text{mol}\times\int_{300K}^{600K}\left(20.9 + 0.42\frac{T'}{\text{K}}\right)d\frac{T'}{\text{K}}\\ &= 20.9\times(600 - 300)\text{J} + \left[0.21\frac{T'}{\text{K}}\right]_{300\text{ K}}^{2\ 600\text{ K}}\text{J}\\ &= 6.27\times 10^3\text{J} + 56.7\times 10^3\text{J}\\ &= 1.19\times 10^4\text{J}\end{aligned}$$

$$\Delta U = \Delta H - \Delta(PV) = \Delta H - nR\Delta T$$
$$= 1.19\times10^4\,\text{J} - 8.314\text{ J K}^{-1}\text{mol}^{-1}\times 300\text{ K}$$
$$= 9.41\times10^3\,\text{J}$$

P2.8) In the adiabatic expansion of 1 mol of an ideal gas from an initial temperature of 25°C, the work done on the surroundings is 1200 J. If $C_{V,m} = 3/2R$, calculate q, w, ΔU, and ΔH.

$q = 0$ because the process is adiabatic

$$\Delta U = w = -1200\text{ J}$$
$$\Delta U = nC_{V,m}\left(T_f - T_i\right)$$
$$T_f = \frac{\Delta U + nC_{V,m}T_i}{nC_{V,m}}$$
$$= \frac{-1200\text{ J} + 7.5\times 8.314\text{ J mol}^{-1}\text{K}^{-1}\times 298\text{ K}}{1.5\times 8.314\text{ J mol}^{-1}\text{K}^{-1}}$$
$$= 202\text{ K}$$
$$\Delta H = nC_{P,m}\left(T_f - T_i\right) = n\left(C_{V,m} + R\right)\left(T_f - T_i\right)$$
$$= 2.5\times 8.314\text{ J mol}^{-1}\text{K}^{-1}\left(202\text{ K} - 298\text{ K}\right)$$
$$= -2.00\times10^3\,\text{J}$$

P2.11) 3.50 moles of an ideal gas is expanded from 450 K and an initial pressure of 5.00 bar to a final pressure of 1.00 bar, and $C_{P,m} = 5/2R$. Calculate w for the following two cases.

a) The expansion is isothermal and reversible.

b) The expansion is adiabatic and reversible.

Without resorting to equations, explain why the result to part (b) is greater than or less than the result to part (a).

a)

$$w = -nRT\ln\frac{V_f}{V_i} = -nRT\ln\frac{P_i}{P_f}$$
$$= -3.50\text{ mol}\times 8.314\text{J mol}^{-1}\text{K}^{-1}\times 450\text{ K}\times\ln\frac{5.00\text{ bar}}{1.00\text{ bar}} = -21.1\times10^3\,\text{J}$$

b) Because $q = 0$, $w = \Delta U$. In order to calculate ΔU, we first calculate T_f.

$$\frac{T_f}{T_i} = \left(\frac{V_f}{V_i}\right)^{1-\gamma} = \left(\frac{T_f}{T_i}\right)^{1-\gamma}\left(\frac{P_i}{P_f}\right)^{1-\gamma}; \left(\frac{T_f}{T_i}\right)^{\gamma} = \left(\frac{P_i}{P_f}\right)^{1-\gamma}; \frac{T_f}{T_i} = \left(\frac{P_i}{P_f}\right)^{\frac{1-\gamma}{\gamma}}$$

$$\frac{T_f}{T_i} = \left(\frac{5.00 \text{ bar}}{1.00 \text{ bar}}\right)^{\frac{1-\frac{5}{3}}{\frac{5}{3}}} = 0.525$$

$$T_f = 0.525 \times 450 \text{ K} = 236 \text{ K}$$

$$w = \Delta U = nC_{V,m}\Delta T = 3.50 \text{ mol} \times \frac{3 \times 8.314 \text{ J mol}^{-1}\text{K}^{-1}}{2} \times (236 \text{ K} - 450 \text{ K}) = -9.34 \times 10^3 \text{J}$$

Less work is done on the surroundings in part (b) because in the adiabatic expansion, the temperature falls and therefore the final volume is less than that in part (a).

P2.13) 3.00 moles of an ideal gas with $C_{V,m} = 3/2R$ initially at a temperature $T_i = 298$ K and $P_i = 1.00$ bar is enclosed in an adiabatic piston and cylinder assembly. The gas is compressed by placing a 625-kg mass on the piston of diameter 20.0 cm. Calculate the work done in this process and the distance that the piston travels. Assume that the mass of the piston is negligible.

$$P_{external} = \frac{F}{A} = \frac{mg}{\pi r^2} = \frac{625 \text{ kg} \times 9.81 \text{m s}^{-2}}{\pi \times (0.100 \text{ m})^2} = 1.95 \times 10^5 \text{ Pa}$$

$$V_i = \frac{nRT}{P_i} = \frac{3.00 \text{ mol} \times 8.314 \text{ J mol}^{-1}\text{K}^{-1} \times 298 \text{ K}}{10^5 \text{ Pa}} = 7.43 \times 10^{-2} \text{m}^3 = 74.3 \text{ L}$$

Following Example Problem 2.6,

$$T_f = T_i\left(\frac{C_{V,m} + \frac{RP_{external}}{P_i}}{C_{V,m} + \frac{RP_{external}}{P_f}}\right) = 298 \text{ K} \times \left(\frac{12.47 \text{ J mol}^{-1}\text{K}^{-1} + \frac{8.314 \text{ J mol}^{-1}\text{K}^{-1} \times 1.95 \times 10^5 \text{ Pa}}{1.00 \times 10^5 \text{ Pa}}}{12.47 \text{ J mol}^{-1}\text{K}^{-1} + \frac{8.314 \text{ J mol}^{-1}\text{K}^{-1} \times 1.95 \times 10^5 \text{ Pa}}{1.95 \times 10^5 \text{ Pa}}}\right)$$

$$= 411 \text{ K}$$

$$V_f = \frac{nRT}{P_f} = \frac{3.00 \text{ mol} \times 8.314 \text{ J mol}^{-1}\text{K}^{-1} \times 411 \text{ K}}{1.95 \times 10^5 \text{Pa}} = 5.25 \times 10^{-2} \text{m}^3$$

$$w = -P_{external}(V_f - V_i) = -1.95 \times 10^5 \text{Pa} \times (5.25 \times 10^{-2} \text{m}^3 - 7.43 \times 10^{-2} \text{m}^3) = 4.25 \times 10^3 \text{J}$$

$$h = \frac{w}{mg} = \frac{4.25 \times 10^3 \text{J}}{625 \text{ kg} \times 9.81 \text{ m s}^{-2}} = 0.69 \text{ m}$$

P2.16) 1 mol of an ideal gas for which $C_{V,m}$ = 20.8 J K^{-1} mol^{-1} is heated from an initial temperature of 0°C to a final temperature of 275°C at constant volume. Calculate *q, w,* ΔU, and ΔH for this process.

$w = 0$ because $\Delta V = 0$

$\Delta U = q = C_V \Delta T = 20.8 \text{ J mol}^{-1}\text{K}^{-1} \times 275 \text{ K} = 5.72\times10^3 \text{J}$

$\Delta H = \Delta U + \Delta(PV) = \Delta U + R\Delta T = 5.72\times10^3\text{J} + 8.314 \text{ J mol}^{-1}\text{K}^{-1} \times 275 \text{ K} = 8.01\times10^3\text{J}$

P2.19) 1 mol of an ideal gas, for which $C_{V,m}$ = 3/2*R*, is subjected to two successive changes in state:
a) From 25.0°C and 100× 10^3 Pa, the gas is expanded isothermally against a constant pressure of 20.0× 10^3 Pa to twice the initial volume.
b) At the end of the previous process, the gas is cooled at constant volume from 25.0°C to –25.0°C. Calculate *q, w,* ΔU, and ΔH for each of the stages. Also calculate *q, w,* ΔU, and ΔH for the complete process.

a) $V_i = \dfrac{nRT}{P_i} = \dfrac{8.314 \text{ J mol}^{-1}\text{K}^{-1} \times 298 \text{ K}}{100R\times10^3 \text{Pa}} = 2.48\times10^{-2}\text{m}^3$

$V_f = 2V_i = 4.96\times10^{-2}\text{m}^3$

$w = -P_{ext}\left(V_f - V_i\right) = -20.0\times10^3\text{Pa}\times\left(4.96\times10^{-2}\text{m}^3 - 2.48\times10^{-2}\text{m}^3\right) = -496 \text{ J}$

ΔU and $\Delta H = 0$ because $\Delta T = 0$

$q = -w = 496 \text{ J}$

b) $\Delta U = nC_{V,m}\left(T_f - T_i\right) = 1.5\times8.314 \text{ J mol}^{-1}\text{K}^{-1}\times\left(248 \text{ K} - 298 \text{ K}\right) = -623 \text{ J}$

$w = 0$ because $\Delta V = 0$

$q = \Delta U = -623 \text{ J}$

$$\Delta H = nC_{P,m}\left(T_f - T_i\right) = n\left(C_{V,m} + R\right)\left(T_f - T_i\right)$$
$$= 2.5\times8.314 \text{ J mol}^{-1}\text{K}^{-1}\times\left(248 \text{ K} - 298 \text{ K}\right)$$
$$= -1.04\times10^3\text{J}$$

$\Delta U_{total} = 0 - 623 \text{ J} = 623 \text{ J}$

$w_{total} = 0 - 496 \text{ J} = -496 \text{ J}$

$q_{total} = 496 \text{ J} - 623 \text{ J} = -127 \text{ J}$

$\Delta H_{total} = 0 - 1.04\times10^3\text{J} = -1.04\times10^3\text{J}$

P2.21) A 1-mol sample of an ideal gas for which $C_{V,m} = 3/2R$ undergoes the following two-step process.
a) From an initial state of the gas described by $T = 28.0°C$ and $P = 2.00 \times 10^4$ Pa, the gas undergoes an isothermal expansion against a constant external pressure of 1.00×10^4 Pa until the volume has doubled.
b) Subsequently, the gas is cooled at constant volume. The temperature falls to –40.5°C.
Calculate q, w, ΔU, and ΔH for each step and for the overall process.

a) For the first step, $\Delta U = \Delta H = 0$ because the process is isothermal.

$$V_i = \frac{nRT_i}{P_i} = \frac{1\ \text{mol} \times 8.314\ \text{J mol}^{-1}\text{K}^{-1} \times (273.15 + 28.0)\ \text{K}}{2.00 \times 10^4 \text{Pa}} = 1.25 \times 10^{-2} \text{m}^3$$

$$w = -q = -P_{external}\Delta V = -1.00 \times 10^4 \text{Pa} \times 0.125 \times 10^{-2} \text{m}^3 = -1.25 \times 10^3 \text{J}$$

b) For the second step, $w = 0$ because $\Delta V = 0$.

$$q = \Delta U = C_V \Delta T = 1\ \text{mol} \times \frac{3 \times 8.314\ \text{J mol}^{-1}\text{K}^{-1}}{2} \times (28.0°\text{C} + 40.5°\text{C}) = 854\ \text{J}$$

$$\Delta H = \Delta U + \Delta(PV) = \Delta U + R\Delta T = 854\ \text{J}\ + 8.314\ \text{J mol}^{-1}\text{K}^{-1} \times (28.0°\text{C} + 40.5°\text{C})$$

$$\Delta H = 1.42 \times 10^3 \text{J}$$

For the overall process, $w = -1.25 \times 10^3$ J, $q = 854 + 1.25 \times 10^3$ J $= 2.02 \times 10^3$ J, $\Delta U = 854$ J, and $\Delta H = 1.42 \times 10^3$ J.

P2.23) A vessel containing 1 mol of an ideal gas with $P_i = 1.00$ bar and $C_{P,m} = 5/2R$ is in thermal contact with a water bath. Treat the vessel, gas, and water bath as being in thermal equilibrium, initially at 298 K, and as separated by adiabatic walls from the rest of the universe. The vessel, gas, and water bath have an average heat capacity of $C_P = 7500\ \text{J K}^{-1}$. The gas is compressed reversibly to $P_f = 10.5$ bar. What is the temperature of the system after thermal equilibrium has been established?

Assume initially that the temperature rise is so small that the reversible compression can be thought of as an isothermal reversible process. If the answer substantiates this assumption, it is valid.

$$w = -nRT_1 \ln\frac{V_f}{V_i} = -nRT_1 \ln\frac{P_i}{P_f}$$

$$= -1\ \text{mol} \times 8.314\ \text{J mol}^{-1}\text{K}^{-1} \times 298\ \text{K} \times \ln\frac{1.00\ \text{bar}}{10.5\ \text{bar}} = 5.83 \times 10^3 \text{J}$$

$$\Delta U_{combined\ system} = C_P \Delta T$$

$$\Delta T = \frac{\Delta U_{combined\ system}}{C_P} = \frac{5.83 \times 10^3 \text{J}}{7500\ \text{J K}^{-1}} = 0.777\ \text{K}$$

$$T_f \approx 299\ \text{K}$$

The result justifies the assumption.

P2.27) Calculate q, w, ΔU, and ΔH if 1.00 mol of an ideal gas with $C_{V,m} = 3/2R$ undergoes a reversible adiabatic expansion from an initial volume $V_i = 5.25$ m^3 to a final volume $V_f = 25.5$ m^3. The initial temperature is 300 K.

$q = 0$ because the process is adiabatic.

$$\frac{T_f}{T_i} = \left(\frac{V_f}{V_i}\right)^{1-\gamma}$$

$$\frac{T_f}{T_i} = \left(\frac{25.5\,L}{5.25L}\right)^{1-\frac{5}{3}} = 0.349$$

$$T_f = 0.349 \times 300 \text{ K} = 105 \text{ K}$$

$$\Delta U = w = nC_{V,n}\Delta T = 1.00 \text{ mol} \times \frac{3 \times 8.314 \text{ J mol}^{-1}\text{K}^{-1}}{2} \times (105 \text{ K} - 300 \text{ K}) = -2.43 \times 10^3 \text{J}$$

$$\Delta H = \Delta U + nR\Delta T = -2.43 \times 10^3 \text{J} + 1.00 \text{ mol} \times 8.314 \text{ J mol}^{-1}\text{K}^{-1} \times (105 \text{ K} - 300 \text{ K})$$

$$\Delta H = -4.05 \times 10^3 \text{ J}$$

P2.29) 1 mol of an ideal gas with $C_{V,m} = 3/2R$ is expanded adiabatically against a constant external pressure of 1.00 bar. The initial temperature and pressure are $T_i = 300$ K and $P_i = 25.0$ bar. The final pressure is $P_f = 1.00$ bar. Calculate q, w, ΔU, and ΔH for the process.

$$\Delta U = nC_{V,m}(T_f - T_i) = -P_{external}(V_f - V_i) = w$$

$q = 0$ because the process is adiabatic.

$$nC_{V,m}(T_f - T_i) = -nRP_{external}\left(\frac{T_f}{P_f} - \frac{T_i}{P_i}\right)$$

$$T_f\left(nC_{V,m} + \frac{nRP_{external}}{P_f}\right) = T_i\left(nC_{V,m} + \frac{nRP_{external}}{P_i}\right)$$

$$T_f = T_i\left(\frac{C_{V,m} + \frac{RP_{external}}{P_i}}{C_{V,m} + \frac{RP_{external}}{P_f}}\right) = 300 \text{ K} \times \left(\frac{1.5 \times 8.314 \text{ J mol}^{-1}\text{K}^{-1} + \frac{8.314 \text{ J mol}^{-1}\text{K}^{-1} \times 1.00 \text{ bar}}{25.0 \text{ bar}}}{1.5 \times 8.314 \text{ J mol}^{-1}\text{K}^{-1} + \frac{8.314 \text{ J mol}^{-1}\text{K}^{-1} \times 1.00 \text{ bar}}{1.00 \text{ bar}}}\right)$$

$$T_f = 185 \text{ K}$$

$$\Delta U = w = nC_{V,n}\Delta T = 1.00 \text{ mol} \times \frac{3 \times 8.314 \text{ J mol}^{-1}\text{K}^{-1}}{2} \times (185 \text{ K} - 300 \text{ K}) = -1.43 \times 10^3 \text{J}$$

$$\Delta H = \Delta U + nR\Delta T = -1.43 \times 10^3 \text{J} + 1.00 \text{ mol} \times 8.314 \text{J mol}^{-1}\text{K}^{-1} \times (185 \text{ K} - 300 \text{ K})$$

$$\Delta H = -2.39 \times 10^3 \text{J}$$

$$100 \times \frac{-5.52 \times 10^3 \text{J} + 5.54 \times 10^3 \text{J}}{-5.52 \times 10^3 \text{J}} = -0.4\%$$

Chapter 3: The Importance of State Functions: Internal Energy and Enthalpy

P3.1) A differential $dz = f(x,y)dx + g(x,y)dy$ is exact if the integral $\int f(x,y)dx + \int g(x,y)dy$ is independent of the path. Demonstrate that the differential $dz = 2xydx + x^2dy$ is exact by integrating dz along the paths $(1,1) \rightarrow (5,1) \rightarrow (5,5)$ and $(1,1) \rightarrow (3,1) \rightarrow (3,3) \rightarrow (5,3) \rightarrow (5,5)$. The first number in each set of parentheses is the x coordinate, and the second number is the y coordinate.

$$\int dz = \int 2xydx + \int x^2dy$$

Path 1

$$\int dz = 2\int_1^5 xdx + 25\int_1^5 dy = \left[x^2\right]_1^5 + 25[y]_1^5 = (25-1) + 25(5-1) = 124$$

Path 2

$$\int dz = 2\int_1^3 xdx + 9\int_1^3 dy + 6\int_3^5 xdx + 25\int_3^5 dy = \left[x^2\right]_1^3 + 9[y]_1^3 + 3\left[x^2\right]_3^5 + 25[y]_3^5$$

$$= (9-1) + 9(3-1) + 3(25-9) + 25(5-3) = 124$$

P3.3) This problem will give you practice in using the cyclic rule. Use the ideal gas law to obtain the three functions $P = f(V,T)$, $V = g(P,T)$, and $T = h(P,V)$. Show that the cyclic rule $\left(\frac{\partial P}{\partial V}\right)_T \left(\frac{\partial V}{\partial T}\right)_P \left(\frac{\partial T}{\partial P}\right)_V = -1$ is obeyed.

$$P = \frac{nRT}{V}; \quad V = \frac{nRT}{P}; \quad T = \frac{PV}{nR}$$

$$\left(\frac{\partial P}{\partial V}\right)_T = -\frac{nRT}{V^2}; \quad \left(\frac{\partial V}{\partial T}\right)_P = \frac{nR}{P}; \quad \left(\frac{\partial T}{\partial P}\right)_V = \frac{V}{nR}$$

$$\left(\frac{\partial P}{\partial V}\right)_T \left(\frac{\partial V}{\partial T}\right)_P \left(\frac{\partial T}{\partial P}\right)_V = \left(-\frac{nRT}{V^2}\right)\left(\frac{nR}{P}\right)\left(\frac{V}{nR}\right) = \frac{-nRT}{PV} = -1$$

P3.6) Because U is a state function, $\left(\frac{\partial}{\partial V}\left(\frac{\partial U}{\partial T}\right)_V\right)_T = \left(\frac{\partial}{\partial T}\left(\frac{\partial U}{\partial V}\right)_T\right)_V$. Using this relationship, show that $\left(\frac{\partial C_V}{\partial V}\right)_T = 0$ for an ideal gas.

For an ideal gas, by definition, $\left(\frac{\partial U}{\partial V}\right)_T = 0$. Because the order of differentiation can be changed for a state function,

$$\left(\frac{\partial}{\partial V}\left(\frac{\partial U}{\partial T}\right)_V\right)_T = \left(\frac{\partial C_V}{\partial V}\right)_T = \left(\frac{\partial}{\partial T}\left(\frac{\partial U}{\partial V}\right)_T\right)_V = 0$$

P3.8) Integrate the expression $\beta = \frac{1}{V}\left(\frac{\partial V}{\partial T}\right)_P$ assuming that β is independent of pressure. By doing so, obtain an expression for V as a function of P and β.

$$\beta = \frac{1}{V}\left(\frac{\partial V}{\partial T}\right)_P$$

$$\frac{dV}{V} = \beta dT$$

$$\int\frac{dV}{V} = \int\beta dT \text{ or } \ln\frac{V_f}{V_i} = \beta\left(T_f - T_i\right)$$

if β can be assumed constant in the temperature interval of interest.

P3.12) Regard the enthalpy as a function of T and P. Use the cyclic rule to obtain the expression $C_P = -\frac{\left(\frac{\partial H}{\partial P}\right)_T}{\left(\frac{\partial T}{\partial P}\right)_H}$.

$$\left(\frac{\partial H}{\partial P}\right)_T\left(\frac{\partial P}{\partial T}\right)_H\left(\frac{\partial T}{\partial H}\right)_P = -1$$

$$C_P = \left(\frac{\partial H}{\partial T}\right)_P = -\left(\frac{\partial H}{\partial P}\right)_T\left(\frac{\partial P}{\partial T}\right)_H = -\frac{\left(\frac{\partial H}{\partial P}\right)_T}{\left(\frac{\partial T}{\partial P}\right)_H}$$

P3.14) Use Equation (3.58), $\left(\frac{\partial U}{\partial V}\right)_T = \frac{\beta T - \kappa P}{\kappa}$ to calculate $\left(\frac{\partial U}{\partial V}\right)_T$ for an ideal gas.

$$\beta = \frac{1}{V}\left(\frac{\partial V}{\partial T}\right)_P = \frac{1}{V}\frac{nR}{P}; \quad \kappa = -\frac{1}{V}\left(\frac{\partial V}{\partial P}\right)_T = \frac{nRT}{VP^2} = \frac{1}{P}$$

$$\left(\frac{\partial U}{\partial V}\right)_T = \frac{\beta T - \kappa P}{\kappa} = \frac{\frac{1}{V}\frac{nRT}{P} - 1}{\frac{1}{P}} = P(1-1) = 0$$

P3.18) Calculate w, q, ΔH, and ΔU for the process in which 1 mol of water undergoes the transition $H_2O(l, 373\text{ K}) \rightarrow H_2O(g, 460\text{ K})$ at 1 bar pressure. The volume of liquid water at 373 K is $1.89 \times 10^{-5}\text{ m}^3\text{ mol}^{-1}$ and the volume of steam at 373 and 460 K is 3.03 and $3.74 \times 10^{-2}\text{ m}^3\text{ mol}^{-1}$, respectively. For steam, $C_{P,m}$ can be considered constant over the temperature interval of interest at 33.58 J $\text{mol}^{-1}\text{ K}^{-1}$.

$$q = \Delta H = n\Delta H_{vaporization} + nC_{P,m}^{steam}\Delta T$$

$$= 40656\text{ J} + 1\text{ mol} \times 33.58\text{ J mol}^{-1}\text{K}^{-1} \times (460\text{ K} - 373\text{ K}) = 4.35 \times 10^4\text{J}$$

$$w = -P_{external}\Delta V = -10^5\text{Pa} \times (3.03 \times 10^{-2}\text{m}^3 - 1.89 \times 10^{-5}\text{m}^3) - 10^5\text{Pa} \times (3.74 \times 10^{-2}\text{m}^3 - 3.03 \times 10^{-2}\text{m}^3)$$

$$= -3028\text{ J} - 710\text{ J} = -3.74 \times 10^3\text{ J}$$

$$\Delta U = w + q = 4.35 \times 10^4\text{J} - 3738\text{ J} = 3.98 \times 10^4\text{J}$$

P3.21) The Joule coefficient is defined as $\left(\frac{\partial T}{\partial V}\right)_U = \frac{1}{C_V}\left[P - T\left(\frac{\partial P}{\partial T}\right)_V\right]$. Calculate the Joule coefficient for an ideal gas and for a van der Waals gas.

For an ideal gas

$$\left(\frac{\partial T}{\partial V}\right)_U = \frac{1}{C_{V,m}}\left[P - T\left(\frac{\partial}{\partial T}\frac{nRT}{V}\right)_V\right] = \frac{1}{C_{V,m}}\left[P - \frac{nRT}{V}\right] = 0$$

For a van der Waals gas

$$\left(\frac{\partial T}{\partial V}\right)_U = \frac{1}{C_{V,m}}\left[P - T\left(\frac{\partial}{\partial T}\left[\frac{RT}{V_m - b} - \frac{a}{V_m^2}\right]\right)_V\right] = \frac{1}{C_{V,m}}\left[P - \frac{RT}{(V_m - b)}\right] = -\frac{1}{C_V}\frac{a}{V_m^2}$$

P3.24) Derive an expression for the internal pressure of a gas that obeys the Bethelot equation of state, $P = \frac{RT}{V_m - b} - \frac{a}{TV_m^2}$.

The internal pressure of a gas is given by

$$\left(\frac{\partial V}{\partial T}\right)_T = T\left(\frac{\partial P}{\partial T}\right)_V - P$$

Using the Bethelot equation of state

$$\left(\frac{\partial P}{\partial T}\right)_V = \frac{R}{V_m - b} + \frac{a}{T^2 V_m^2}$$

$$\left(\frac{\partial U}{\partial V}\right)_T = \frac{RT}{V_m - b} + \frac{a}{TV_m^2} - \left(\frac{RT}{V_m - b} - \frac{a}{TV_m^2}\right) = \frac{2a}{TV_m^2}$$

P3.27) Use the result of the Problem P3.26 to show that $\left(\frac{\partial C_V}{\partial V}\right)_T$ for the van der Waals gas is zero.

We use the relationship

$$\left(\frac{\partial C_V}{\partial V}\right)_T = T\left(\frac{\partial^2 P}{\partial T^2}\right)_V$$

$$P = \frac{RT}{V_m - b} - \frac{a}{V_m^2}$$

$$\left(\frac{\partial P}{\partial T}\right)_V = \frac{R}{V_m - b}$$

$$\left(\frac{\partial^2 P}{\partial T^2}\right)_V = \left(\frac{\partial \frac{R}{V_m - b}}{\partial T}\right)_V = 0$$

therefore $\left(\frac{\partial C_V}{\partial V}\right)_T = T\left(\frac{\partial^2 P}{\partial T^2}\right)_V = T \times 0 = 0$

P3.30) Use the relation $C_{P,m} - C_{V,m} = T\left(\frac{\partial V_m}{\partial T}\right)_P\left(\frac{\partial P}{\partial T}\right)_P$, the cyclic rule, and the van der Waals equation of state to derive an equation for $C_{P,m} - C_{V,m}$ in terms of V_m and the gas constants R, a, and b.

We use the cyclic rule to evaluate $\left(\frac{\partial V_m}{\partial T}\right)_P$.

$$\left(\frac{\partial V_m}{\partial T}\right)_P \left(\frac{\partial T}{\partial P}\right)_{V_m} \left(\frac{\partial P}{\partial V_m}\right)_T = -1$$

$$\left(\frac{\partial V_m}{\partial T}\right)_P = -\left(\frac{\partial P}{\partial T}\right)_{V_m} \left(\frac{\partial V_m}{\partial P}\right)_T$$

$$C_{P,m} - C_{V,m} = T\left(\frac{\partial V_m}{\partial T}\right)_P \left(\frac{\partial P}{\partial T}\right)_{V_m} = -T\left[\left(\frac{\partial P}{\partial T}\right)_{V_m}\right]^2 \left(\frac{\partial V_m}{\partial P}\right)_T = -T\frac{\left[\left(\frac{\partial P}{\partial T}\right)_{V_m}\right]^2}{\left(\frac{\partial P}{\partial V_m}\right)_T}$$

$$P = \frac{RT}{V_m - b} - \frac{a}{V_m^2}$$

$$\left(\frac{\partial P}{\partial T}\right)_{V_m} = \frac{R}{V_m - b}$$

$$\left(\frac{\partial P}{\partial V_m}\right)_T = \frac{-RT}{(V_m - b)^2} + \frac{2a}{V_m^3} = \frac{-RTV_m^3 + 2a(V_m - b)}{V_m^3 (V_m - b)^2}$$

$$C_{P,m} - C_{V,m} = -T\frac{\left(\frac{R}{V_m - b}\right)^2}{\frac{-RT}{(V_m - b)^2} + \frac{2a}{V_m^3}} = -T\frac{R}{-T + \frac{2a(V_m - b)^2}{RV_m^3}} = \frac{R}{1 - \frac{2a(V_m - b)^2}{RTV_m^3}}$$

In the ideal gas limit, $a = 0$, and $C_{P,m} - C_{V,m} = R$.

P3.32) Show that the expression $\left(\frac{\partial U}{\partial V}\right)_T = T\left(\frac{\partial P}{\partial T}\right)_V - P$ can be written in the form

$$\left(\frac{\partial U}{\partial V}\right)_T = T^2\left(\frac{\partial [P/T]}{\partial T}\right)_V = -\left(\frac{\partial [P/T]}{\partial [1/T]}\right)_V.$$

$$\left(\frac{\partial U}{\partial V}\right)_T = T\left(\frac{\partial P}{\partial T}\right)_V - P$$

$$\left(\frac{\partial [P/T]}{\partial T}\right)_V = P\left(\frac{\partial [1/T]}{\partial T}\right)_V + \frac{1}{T}\left(\frac{\partial P}{\partial T}\right)_V$$

$$= -\frac{P}{T^2} + \frac{1}{T}\left(\frac{\partial P}{\partial T}\right)_V$$

$$\left(\frac{\partial P}{\partial T}\right)_V = T\left(\left(\frac{\partial [P/T]}{\partial T}\right)_V + \frac{P}{T^2}\right)$$

$$\left(\frac{\partial U}{\partial V}\right)_T = T^2\left(\left(\frac{\partial [P/T]}{\partial T}\right)_V + \frac{P}{T^2}\right) - P$$

$$= T^2\left(\frac{\partial [P/T]}{\partial T}\right)_V + P - P = T^2\left(\frac{\partial [P/T]}{\partial T}\right)_V$$

We now change the differentiation to the variable $1/T$.

$$\left(\frac{\partial [P/T]}{\partial T}\right)_V = \left(\frac{\partial [P/T]}{\partial [1/T]}\right)_V \left(\frac{\partial [1/T]}{\partial T}\right)_V = -\frac{1}{T^2}\left(\frac{\partial [P/T]}{\partial [1/T]}\right)_V$$

$$\left(\frac{\partial U}{\partial V}\right)_T = T^2\left(\frac{\partial [P/T]}{\partial T}\right)_V = T^2\left(-\frac{1}{T^2}\frac{\partial [P/T]}{\partial [1/T]}\right)_V = -\left(\frac{\partial [P/T]}{\partial [1/T]}\right)_V$$

Chapter 4: Thermochemistry

P4.2) Calculate $\Delta H^\circ_{reaction}$ and $\Delta U^\circ_{reaction}$ for the oxidation of benzene. Also calculate $\dfrac{\Delta H^\circ_{reaction} - \Delta U^\circ_{reaction}}{\Delta H^\circ_{reaction}}$.

$15/2\ O_2(g) + C_6H_6(l) \rightarrow 3H_2O(l) + 6CO_2(g)$

From the data tables,

$$\begin{aligned}\Delta H^\circ_{combustion} &= 3\Delta H^\circ_f(\mathrm{H_2O},l) + 6\Delta H^\circ_f(\mathrm{CO_2},g) - \Delta H^\circ_f(\mathrm{C_6H_6},l)\\ &= -3\times 285.8\,\mathrm{kJ\,mol^{-1}} - 6\times 393.5\,\mathrm{kJ\,mol^{-1}} - 49.1\,\mathrm{kJ\,mol^{-1}}\\ &= 3268\,\mathrm{kJ\,mol^{-1}}\end{aligned}$$

$$\begin{aligned}\Delta U^\circ_{reaction} &= \Delta H^\circ_{reaction} - \Delta nRT = -3268\,\mathrm{kJ\,mol^{-1}} + 1.5\times 8.314\,\mathrm{J\,K^{-1}mol^{-1}}\times 298.15\ \mathrm{K}\\ &= -3264\,\mathrm{kJ\,mol^{-1}}\end{aligned}$$

$$\frac{\Delta H^\circ_{reaction} - \Delta U^\circ_{reaction}}{\Delta H^\circ_{reaction}} = \frac{-3268\,\mathrm{kJ\,mol^{-1}} + 3264\,\mathrm{kJ\,mol^{-1}}}{-3268\,\mathrm{kJ\,mol^{-1}}} = 0.0122$$

P4.4) Calculate ΔH°_f for $N_2(g)$ at 650 K using the temperature dependence of the heat capacities from the data tables. How large is the relative error if the molar heat capacity is assumed to be constant at its value of 298.15 K over the temperature interval?

$$\begin{aligned}\Delta H^\circ_f(\mathrm{N_2}, g, 650\ \mathrm{K}) &= \Delta H^\circ_f(\mathrm{N_2}, g, 298.15\ \mathrm{K}) + \int_{298.15}^{650} C_{P,m}\left(\frac{T}{\mathrm{K}}\right) d\frac{T}{\mathrm{K}}\\ &= \left[\int_{298.15}^{650}\left(30.81 - 0.01187\frac{T}{\mathrm{K}} + 2.3968\times10^{-5}\frac{T^2}{\mathrm{K}^2} - 1.0176\times10^{-8}\frac{T^3}{\mathrm{K}^3}\right) d\frac{T}{\mathrm{K}}\right]\mathrm{J\,K^{-1}mol^{-1}}\\ &= (10841 - 1980 + 1982 - 434.0)\,\mathrm{J\,mol^{-1}} = 10.41\ \mathrm{kJ\,mol^{-1}}\end{aligned}$$

If it is assumed that the heat capacity is constant at its value at 298 K,

$$\Delta H^\circ_f \approx \left[\int_{298.15}^{650}(29.13)\,d\frac{T}{\mathrm{K}}\right]\mathrm{J\,K^{-1}mol^{-1}} = 10.25\ \mathrm{kJ\,mol^{-1}}$$

$$\text{Error} = 100\times\frac{10.25\ \mathrm{kJ\,mol^{-1}} - 10.41\ \mathrm{kJ\,mol^{-1}}}{10.41\ \mathrm{kJ\,mol^{-1}}} = -1.54\%$$

P4.9) From the following data at 298.15 K as well as Data Table 4.1 (Appendix B, Data Tables), calculate the standard enthalpy of formation of $H_2S(g)$ and of $FeS_2\ (s)$:

	$\Delta H^\circ_{reaction}$ (kJ mol^{-1})
$Fe(s) + 2H_2S(g) \rightarrow FeS_2(s) + 2H_2(g)$	–137.0
$H_2S(g) + 3/2\ O_2(g) \rightarrow H_2O(l) + SO_2(g)$	–562.0

	$\Delta H^\circ_{reaction}$ (kJ mol^{-1})
$H_2O(l) + SO_2(g) \rightarrow H_2S(g) + 3/2O_2(g)$	562.0
$S(s) + O_2(g) \rightarrow SO_2(g)$	–296.8
$H_2(g) + 1/2O_2(g) \rightarrow H_2O(l)$	–285.8
$H_2(g) + S(s) \rightarrow H_2S(g)$	$\Delta H^\circ_f = -20.6$ kJ mol^{-1}

	$\Delta H^\circ_{reaction}$ (kJ mol^{-1})
$Fe(s) + 2H_2S(g) \rightarrow FeS_2(s) + 2H_2(g)$	–137.0
$2H_2(g) + 2S(s) \rightarrow 2H_2S(g)$	–2 × 20.6
$Fe(s) + 2S(s) \rightarrow FeS_2(s)$	$\Delta H^\circ_f = -178.2$ kJ mol^{-1}

P4.11) Use the average bond energies in Table 4.3 to estimate ΔU for the reaction $C_2H_4(g) + H_2(g) \rightarrow C_2H_6(g)$. Also calculate $\Delta U^\circ_{reaction}$ from the tabulated values of ΔH°_f for reactant and products (Appendix B, Data Tables). Calculate the percent error in estimating $\Delta U^\circ_{reaction}$ from the average bond energies for this reaction.

$\Delta U_{reaction}$ = –(C–C bond energy + 6 C–H bond energy – H–H bond energy – C=C bond energy – 4 C–H bond energy)
$\Delta U_{reaction}$ = –(346 kJ mol^{-1} + 6 × 411 kJ mol^{-1} – 432 kJ mol^{-1} – 602 kJ mol^{-1} – 4 × 411 kJ mol^{-1}) = –134 kJ mol^{-1}.

Using the data tables,

$$\Delta H^\circ_{reaction}(298.15\text{ K}) = \Delta H^\circ_f(C_2H_6,g) - \Delta H^\circ_f(C_2H_4,g) - \Delta H^\circ_f(H_2,g)$$

$$\Delta H^\circ_{reaction}(298.15\text{ K}) = -84.0\text{ kJ mol}^{-1} - 52.4\text{ kJ mol}^{-1} = -136.4\text{ kJ mol}^{-1}$$

$$\Delta U^\circ_{reaction}(298.15\text{ K}) = \Delta H^\circ_{reaction}(298.15\text{ K}) - \Delta nRT$$

$$= -136.4\text{ kJ mol}^{-1} + 8.314\text{ J mol}^{-1}\text{ K}^{-1} \times 298.15\text{ K} = -133.9\text{ kJ mol}^{-1}$$

$$\text{Relative Error} = 100 \times \frac{+134\text{ kJ mol}^{-1} - 133.9\text{ kJ mol}^{-1}}{-133.9\text{ kJ mol}^{-1}} \approx 0\%$$

P4.14) At 298 K, $\Delta H^\circ_{reaction}$ = 131.28 kJ mol^{-1} for the reaction

C(*graphite*) + $H_2O(g) \rightarrow CO(g) + H_2(g)$, with $C_{P,m}$ = 8.53, 33.58, 29.12, and 28.82 J K^{-1} mol^{-1} for graphite, $H_2O(g)$, $CO(g)$, and $H_2(g)$, respectively. Calculate $\Delta H^\circ_{reaction}$ at 125°C from this information. Assume that the heat capacities are independent of temperature.

$$\Delta H^\circ_{reaction}(398\text{ K}) = \Delta H^\circ_{reaction}(298\text{ K}) + \int_{298\text{ K}}^{398\text{ K}} \Delta C_P(T)\,dT$$

For this problem, it is assumed that the heat capacities are independent of *T*.

$$\begin{aligned}
&\Delta H^\circ_{reaction}(398\text{ K}) = \Delta H^\circ_{reaction}(298\text{ K}) \\
&+\left[C_{P,m}(\text{H}_2,g)+C_{P,m}(\text{CO},g)-C_{P,m}(\text{C},graphite)-C_{P,m}(\text{H}_2\text{O},(g))\right]\Delta T \\
&=131.28\text{ kJ mol}^{-1}+[28.82+29.12-8.53-33.58]\text{J mol}^{-1}\text{ K}^{-1}\times 100\text{ K} \\
&=132.86\text{ kJ mol}^{-1}
\end{aligned}$$

P4.16) Consider the reaction
$TiO_2(s)$ + 2 C(*graphite*) + 2 $Cl_2(g) \rightarrow$ 2 CO(*g*) + $TiCl_4(l)$ for which
$\Delta H^\circ_{reaction,\,298\text{K}} = -80.01$ kJ mol^{-1}. Given the following data at 25°C,

Substance	$TiO_2(s)$	$Cl_2(g)$	C(*graphite*)	CO(*g*)	$TiCl_4(l)$
ΔH°_f (kJ mol^{-1})	–945			–110.5	
$C_{P,m}$ (J K^{-1} mol^{-1})	55.06	33.91	8.53	29.12	145.2

a) calculate $\Delta H^\circ_{reaction}$ at 135.8°C, the boiling point of $TiCl_4$, and
b) calculate ΔH°_f for $TiCl_4$ (*l*) at 25°C.
Assume that the heat capacities are independent of temperature.

a) $$\Delta H^\circ_{reaction}(409.0\text{ K}) = \Delta H^\circ_{reaction}(298\text{ K}) + \int_{298\text{ K}}^{409.0\text{ K}} \Delta C_{P,m}\,dT$$

In this case, the heat capacities are assumed to be independent of *T*.

$$\begin{aligned}
&\Delta H^\circ_{reaction}(409.0\text{ K}) = \Delta H^\circ_{reaction}(298\text{ K}) + \Delta C_{P,m}[409.0\text{ K}-298\text{ K}] \\
&\quad=-80.1\text{ kJ mol}^{-1}+\left[C_{P,m}(\text{TiCl}_4,l)+2C_{P,m}(\text{CO},g)-C_{P,m}(\text{TiO}_2,s)\right. \\
&\qquad\qquad\left.-2C_{P,m}(graphite,s)-2C_{P,m}(\text{Cl}_2,g)\right][409.0\text{ K}-298\text{ K}] \\
&\quad=-80.01\text{ kJ mol}^{-1}+[145.2+2\times29.12-55.06-2\times8.53-2\times33.91][409.0\text{ K}-298\text{ K}] \\
&\quad= -73.0\text{ kJ mol}^{-1}
\end{aligned}$$

b) $TiO_2(s)$ + 2C(*graphite*) + $2Cl_2(g) \rightarrow 2CO(g) + TiCl_4(l)$

$$\Delta H^\circ_{reaction} = -80.01 \text{ kJ mol}^{-1} = 2\Delta H^\circ_f(\text{CO},g) + \Delta H^\circ_f(\text{TiCl}_4,l) - \Delta H^\circ_f(\text{TiO}_2,s)$$

$$\Delta H^\circ_f(\text{TiCl}_4,l) = \Delta H^\circ_f(\text{TiO}_2,s) - 2\Delta H^\circ_f(\text{CO},g) - 80.01 \text{ kJ mol}^{-1}$$

$$= -945 \text{ kJ mol}^{-1} + 2\times 110.5 \text{ kJ mol}^{-1} - 80.01 \text{ kJ mol}^{-1}$$

$$= -804 \text{ kJ mol}^{-1}$$

P4.20) A sample of K(*s*) of mass 2.140 g undergoes combustion in a constant volume calorimeter. The calorimeter constant is 1849 J K^{-1}, and the measured temperature rise in the inner water bath containing 1450 g of water is 2.62 K. Calculate ΔU°_f and ΔH°_f for K_2O.

2K(*s*) + ½ O_2(g) → K_2O(*s*)

$$\Delta U_f = -\frac{M_s}{m_s}\left(\frac{m_{H_2O}}{M_{H_2O}} C_{H_2O,m}\Delta T + C_{calorimeter}\Delta T\right)$$

$$= -\frac{39.098 \text{ g mol}^{-1}}{2.240 \text{ g}} \times \frac{2 \text{ mol K}}{1 \text{ mol reaction}} \times \left(\begin{array}{l} \dfrac{1.450 \times 10^3 \text{g}}{18.02 \text{ g mol}^{-1}} \times 75.291 \text{J mol}^{-1}\text{ K}^{-1} \times 2.62^\circ\text{C} \\ +1.849\times 10^3 \text{J}^\circ\text{C}^{-1} \times 2.62^\circ\text{C} \end{array}\right)$$

$$= -757 \text{ kJ mol}^{-1}$$

$$\Delta H^\circ_f = \Delta U^\circ_f + \Delta nRT$$

$$= -757 \text{ kJ mol}^{-1} + \frac{1}{2} \times 8.3145 \text{J K}^{-1}\text{mol}^{-1} \times 298.15 \text{ K} = -756 \text{ kJ mol}^{-1}$$

$$\frac{-2.86 \text{ kJ mol}^{-1} + 2.4 \text{ kJ mol}^{-1}}{-2.86 \text{ kJ mol}^{-1}} = 16\%$$

Chapter 5: Entropy and the Second and Third Laws of Thermodynamics

P5.2) Consider the reversible Carnot cycle shown in Figure 5.2 with 1 mol of an ideal gas with $C_V = 3/2R$ as the working substance. The initial isothermal expansion occurs at the hot reservoir temperature of $T_{hot} = 600°C$ from an initial volume of 3.50 L (V_a) to a volume of 10.0 L (V_b). The system then undergoes an adiabatic expansion until the temperature falls to $T_{cold} = 150°C$. The system then undergoes an isothermal compression and a subsequent adiabatic compression until the initial state described by $T_a = 600°C$ and $V_a = 3.50$ L is reached.

a) Calculate V_c and V_d.
b) Calculate w for each step in the cycle and for the total cycle.
c) Calculate ε and the amount of heat that is extracted from the hot reservoir to do 1.00 kJ of work in the surroundings.

a) We first calculate V_c and V_d.

$$\frac{T_c}{T_b} = \left(\frac{V_c}{V_b}\right)^{1-\gamma}; \frac{V_c}{V_b} = \left(\frac{T_c}{T_b}\right)^{\frac{1}{1-\gamma}}$$

$$\frac{V_c}{V_b} = \left(\frac{423\text{ K}}{873\text{ K}}\right)^{\frac{1}{1-\frac{5}{3}}} = \left(\frac{423\text{ K}}{873\text{ K}}\right)^{-\frac{3}{2}} = 2.96; \quad V_c = 29.6\text{ L}$$

$$\frac{T_d}{T_a} = \left(\frac{V_d}{V_a}\right)^{1-\gamma}; \frac{V_d}{V_a} = \left(\frac{T_d}{T_a}\right)^{\frac{1}{1-\gamma}}$$

$$\frac{V_d}{V_a} = \left(\frac{423\text{ K}}{873\text{ K}}\right)^{\frac{1}{1-\frac{5}{3}}} = \left(\frac{423\text{ K}}{873\text{ K}}\right)^{-\frac{3}{2}} = 2.96; \quad V_d = 10.4\text{ L}$$

b) We next calculate w for each step in the cycle, and for the total cycle.

$$w_{ab} = -nRT_a \ln\frac{V_b}{V_a} = -1\text{ mol}\times 8.314\text{ J mol}^{-1}\text{K}^{-1}\times 873\text{ K}\times\ln\frac{10.0\text{ L}}{3.50\text{ L}} = -7.62\times10^3\text{J}$$

$$w_{bc} = nC_{V,m}\left(T_c - T_b\right) = 1\text{ mol}\times\frac{3}{2}\times 8.314\text{ J mol}^{-1}\text{K}^{-1}\times\left(423\text{ K} - 873\text{ K}\right) = -5.61\times10^3\text{J}$$

$$w_{cd} = -nRT_c \ln\frac{V_d}{V_c} = -1\text{ mol}\times 8.314\text{ J mol}^{-1}\text{K}^{-1}\times 423\text{ K}\times\ln\frac{10.4\text{ L}}{29.6\text{ L}} = +3.68\times10^3\text{J}$$

$$w_{da} = nC_{V,m}\left(T_a - T_d\right) = 1\text{ mol}\times\frac{3}{2}\times 8.314\text{ J mol}^{-1}\text{K}^{-1}\times\left(873\text{ K} - 423\text{ K}\right) = +5.61\times10^3\text{J}$$

$$w_{total} = -7.62\times10^3\text{J} - 5.61\times10^3\text{J} + 3.68\times10^3\text{J} + 5.61\times10^3\text{J} = -3.94\times10^3\text{J}$$

$$\varepsilon = 1 - \frac{T_{cold}}{T_{hot}} = 1 - \frac{423\text{ K}}{873\text{ K}} = 0.515 \quad q = \frac{w}{\varepsilon} = 1.94$$

Therefore, 1.94 kJ of heat must be extracted from the surroundings to do 1.00 kJ of work in the surroundings.

P5.5) Calculate ΔS if the temperature of 1 mol of an ideal gas with $C_V = 3/2R$ is increased from 150 to 350 K under conditions of (a) constant pressure and (b) constant volume.

a) at constant pressure

$$\Delta S = nC_{P,m} \ln\frac{T_f}{T_i} = 1\text{ mol}\times\left(\frac{3}{2}+1\right)\times 8.314\text{ J mol}^{-1}\text{K}^{-1}\times\ln\frac{350\text{ K}}{150\text{ K}} = 17.6\text{ J K}^{-1}$$

b) at constant volume

$$\Delta S = nC_{V,m} \ln\frac{T_f}{T_i} = 1\text{ mol}\times\frac{3}{2}\times 8.314\text{ J mol}^{-1}\text{K}^{-1}\times\ln\frac{350\text{ K}}{150\text{ K}} = 10.6\text{ J K}^{-1}$$

P5.9) At the transition temperature of 95.4°C, the enthalpy of transition from rhombic to monoclinic sulfur is 0.38 kJ mol^{-1}.

a) Calculate the entropy of transition under these conditions.

b) At its melting point, 119°C, the enthalpy of fusion of monoclinic sulfur is 1.23 kJ mol^{-1}. Calculate the entropy of fusion.

c) The values given in parts (a) and (b) are for 1 mol of sulfur; however, in crystalline and liquid sulfur, the molecule is present as S_8. Convert the values of the enthalpy and entropy of fusion in parts (a) and (b) to those appropriate for S_8.

a) $$\Delta S_{transition} = \frac{\Delta H_{transition}}{T_{transition}} = \frac{0.38\text{ kJ mol}^{-1}}{(273.15+95.4)\text{ K}} = 1.03\text{ J K}^{-1}\text{ mol}^{-1}$$

b) $$\Delta S_{fusion} = \frac{\Delta H_{fusion}}{T_{fusion}} = \frac{1.23\text{ kJ mol}^{-1}}{(273.15+119)\text{ K}} = 3.14\text{ J K}^{-1}\text{ mol}^{-1}$$

c) Each of the ΔS in parts (a) and (b) should be multiplied by 8.

$$\Delta S_{transition} = 8.24\text{ J K}^{-1}\text{ mol}^{-1}$$
$$\Delta S_{fusion} = 25.12\text{ J K}^{-1}\text{ mol}^{-1}$$

P5.10)

a) Calculate ΔS if 1 mol of liquid water is heated from 0° to 100°C under constant pressure if $C_{P,m} = 75.291$ J K^{-1} mol^{-1}.

b) The melting point of water at the pressure of interest is 0°C and the enthalpy of fusion is 6.0095 kJ mol^{-1}. The boiling point is 100°C and the enthalpy of vaporization is 40.6563 kJ mol^{-1}. Calculate ΔS for the transformation $H_2O(s, 0°C) \rightarrow H_2O(g, 100°C)$.

a) The heat input is the same for a reversible and an irreversible process.
$dq = dq_{reversible} = nC_{P,m}dT$

$$\Delta S = n\int \frac{C_{P,m}}{T}dT = nC_{P,m}\ln\frac{T_f}{T_i}$$

$$= 1\text{ mol} \times 753\text{ J mol}^{-1}\text{ K}^{-1}\ln\frac{373.15\text{ K}}{273.15\text{ K}}$$

$$= 23.49\text{ J K}^{-1}$$

b)

$$\Delta S_{fusion} = \frac{\Delta H_{fusion}}{T_{fusion}} = \frac{6009\text{ J}}{273.15\text{ K}} = 22.00\text{ J K}^{-1}$$

$$\Delta S_{vaporization} = \frac{\Delta H_{vaporization}}{T_{vaporization}} = \frac{40656\text{ J}}{373.15\text{ K}} = 108.95\text{ J K}^{-1}$$

$$\Delta S_{total} = \Delta S_{fusion} + \Delta S_{vaporization} + \Delta S_{heating} = (22.00 + 108.95 + 23.49)\text{ J K}^{-1}$$

$$= 154.4\text{ J K}^{-1}$$

P5.13) Between 0°C and 100°C, the heat capacity of Hg(l) is given by $\frac{C_{P,m}(\text{Hg},l)}{\text{J K}^{-1}\text{ mol}^{-1}} = 30.093 - 4.944\times10^{-3}\frac{T}{\text{K}}$. Calculate ΔH and ΔS if 1 mol of Hg(l) is raised in temperature from 0° to 100°C at constant P.

$$\Delta H_m = \int_{T_i}^{T_f} C_{P,m}d[T/K]$$

$$= 30.093(T_f - T_i) - 2.472\times10^{-3}(T_f^2 - T_i^2)\text{ J mol}^{-1}$$

$$= 2.84\times10^3\text{ J mol}^{-1}$$

$$\Delta S_m = \int_{T_i}^{T_f} \frac{C_{P,m}}{[T/K]}d[T/K] = 30.093\ln\frac{T_f}{T_i}$$

$$-4.944\times10^{-3}(T_f - T_i) = 8.90\text{ J K}^{-1}\text{ mol}^{-1}$$

P5.15) The heat capacity of α-quartz is given by $\frac{C_{P,m}(\alpha\text{-quartz},s)}{\text{J K}^{-1}\text{ mol}^{-1}} = 46.94 + 34.31\times10^{-3}\frac{T}{\text{K}} - 11.30\times10^{-5}\frac{T^2}{\text{K}^2}$. The coefficient of thermal expansion is given by $\beta = 0.3530 \times 10^{-4}\text{ K}^{-1}$ and $V_m = 22.6\text{ cm}^3\text{ mol}^{-1}$. Calculate ΔS_m for the transformation α-quartz (25°C, 1 atm) → α-quartz (225°C, 1000 atm).

From Equations (5.23) and (5.24)

$$\Delta S = \int_{T_i}^{T_f} C_P \frac{dT}{T} - V\beta(P_f - P_i)$$

$$= \left[46.94\ln\frac{498\text{ K}}{298\text{ K}} + 34.31\times10^{-3}\times(498-298) - 5.65\times10^{-5}\times(498^2-298^2)\right]\text{J K}^{-1}\text{ mol}^{-1}$$

$$-22.6\text{ cm}^3\text{ mol}^{-1}\times\frac{1\text{ m}^3}{10^6\text{ cm}^3}\times0.3530\times10^{-4}\text{ K}^{-1}\times999\text{ atm}\times\frac{1.0125\times10^5\text{Pa}}{\text{atm}}$$

$$= 22.11\text{ J K}^{-1}\text{ mol}^{-1} - 0.0807\text{ J K}^{-1}\text{ mol}^{-1} = 21.88\text{ J K}^{-1}\text{ mol}^{-1}$$

P5.17) Calculate ΔS, ΔS_{total}, and $\Delta S_{surroundings}$ when the volume of 85.0 g of CO initially at 298 K and 1.00 bar increases by a factor of three in (a) an adiabatic reversible expansion, (b) an expansion against $P_{external} = 0$, and (c) an isothermal reversible expansion. Take $C_{P,m}$ to be constant at the value 29.14 J mol^{-1} K^{-1} and assume ideal gas behavior. State whether each process is spontaneous.

a) an adiabatic reversible expansion
$\Delta S_{surroundings} = 0$ because $q = 0$. $\Delta S = 0$ because the process is reversible.

$\Delta S_{total} = \Delta S + \Delta S_{surroundings} = 0$. The process is not spontaneous.

b) an expansion against $P_{external} = 0$
ΔT and $w = 0$. Therefore $\Delta U = q = 0$.

$$\Delta S = nR\ln\frac{V_f}{V_i} = \frac{85.0\text{ g}}{28.01\text{ g mol}^{-1}}\times8.314\text{ J mol}^{-1}\text{K}^{-1}\times\ln 3$$

$$= 3.03\text{ mol}\times9.13\text{ J mol}^{-1}\text{K}^{-1} = 27.7\text{ J K}^{-1}$$

$\Delta S_{total} = \Delta S + \Delta S_{surroundings} = 27.7\text{ J K}^{-1} + 0 = 27.17\text{ J K}^{-1}$. The process is spontaneous.

c) an isothermal reversible expansion
$\Delta T = 0$. Therefore $\Delta U = 0$.

$$w = -q = -nRT\ln\frac{V_f}{V_i} = -3.03\text{ mol}\times8.314\text{ J mol}^{-1}\text{K}^{-1}\times298\text{ K}\times\ln 3 = -8.25\times10^3\text{J}$$

$$\Delta S = \frac{q_{reversible}}{T} = \frac{8.25\times10^3\text{J}}{298\text{ K}} = 27.7\text{ J K}^{-1}$$

$$\Delta S_{surroundings} = \frac{-q}{T} = \frac{-8.25\times10^3\text{J}}{298\text{ K}} = -27.7\text{ J K}^{-1}$$

$\Delta S_{total} = \Delta S + \Delta S_{surroundings} = 27.7\text{ J K}^{-1} - 27.7\text{ J K}^{-1} = 0$. The system and surroundings are at equilibrium.

P5.20) One mole of $H_2O(l)$ is compressed from a state described by $P = 1.00$ bar and $T = 298$ K to a state described by $P = 800$ bar and $T = 450$ K. In addition, $\beta = 2.07\times10^{-4}$ K^{-1} and the density can be assumed to be constant at the value 997 kg m^{-3}. Calculate ΔS for this transformation, assuming that $\kappa = 0$.

From Equation (5.24),

$$\Delta S = \int_{T_i}^{T_f} \frac{C_P}{T} dT - \int_{P_i}^{P_f} V \beta dP \approx nC_{P,m} \ln \frac{T_f}{T_i} - nV_{m,i}\beta\left(P_f - P_i\right)$$

$$= 1\,\text{mol} \times 75.3\ \text{J mol}^{-1}\text{K}^{-1} \times \ln \frac{450\ \text{K}}{298\ \text{K}}$$

$$-1\,\text{mol} \times \frac{18.02 \times 10^{-3}\text{kg mol}^{-1}}{997\ \text{kg m}^{-3}} \times 2.07 \times 10^{-4}\text{K}^{-1} \times 799\ \text{bar} \times 10^{5}\text{Pa bar}^{-1}$$

$$= 31.0\,\text{J K}^{-1} - 0.299\ \text{J K}^{-1} = 30.7\,\text{J K}^{-1}$$

P5.23) The maximum theoretical efficiency of an internal combustion engine is achieved in a reversible Carnot cycle. Assume that the engine is operating in the Otto cycle and that $C_{V,m} = 5/2R$ for the fuel–air mixture initially at 298 K (the temperature of the cold reservoir). The mixture is compressed by a factor of 8.0 in the adiabatic compression step. What is the maximum theoretical efficiency of this engine? How much would the efficiency increase if the compression ratio could be increased to 30? Do you see a problem in doing so?

$$\frac{T_f}{T_i} = \left(\frac{V_f}{V_i}\right)^{1-\gamma} = \left(\frac{1}{8}\right)^{1-\frac{7}{5}} = \left(\frac{1}{8}\right)^{-0.4} = 2.30$$

$$T_f = 2.30 \times 298\ \text{K} = 684\ \text{K}$$

$$\varepsilon = 1 - \frac{T_{cold}}{T_{hot}} = 1 - \frac{298}{684} = 0.564$$

$$\frac{T_f}{T_i} = \left(\frac{V_f}{V_i}\right)^{1-\gamma} = \left(\frac{1}{30}\right)^{1-\frac{7}{5}} = \left(\frac{1}{30}\right)^{-0.4} = 3.89$$

$$T_f = 3.89 \times 298\ \text{K} = 1162\ \text{K}$$

$$\varepsilon = 1 - \frac{T_{cold}}{T_{hot}} = 1 - \frac{298\ \text{K}}{1162\ \text{K}} = 0.744$$

It would be difficult to avoid ignition of the fuel-air mixture before the compression was complete.

P5.26) The interior of a refrigerator is typically held at 277 K and the interior of a freezer is typically held at 255 K. If the room temperature is 294 K, by what factor is it more expensive to extract the same amount of heat from the freezer than from the refrigerator? Assume that the theoretical limit for the performance of a reversible refrigerator is valid.

From Equation (5.44)

$$\eta_r = \frac{T_{cold}}{T_{hot} - T_{cold}}$$

for the freezer $\eta_r = \dfrac{255\text{ K}}{294\text{ K} - 255\text{ K}} = 6.5$

for the refrigerator $\eta_r = \dfrac{277\text{ K}}{294\text{ K} - 277\text{ K}} = 16.3$

The freezer is more expensive to operate than the refrigerator by the ratio 16.3/6.5 = 2.5.

P5.28) The mean solar flux at the Earth's surface is ~4.00 J cm^{-2} min^{-1}. In a nonfocusing solar collector, the temperature can reach a value of 90.0°C. A heat engine is operated using the collector as the hot reservoir and a cold reservoir at 298 K. Calculate the area of the collector needed to produce one horsepower (1 hp = 746 watts). Assume that the engine operates at the maximum Carnot efficiency.

$$\varepsilon = 1 - \frac{T_{hot}}{T_{cold}} = 1 - \frac{298\text{ K}}{363\text{ K}} = 0.179$$

The area required for the solar panel is $\dfrac{2.5\times10^{5}\text{ J min}^{-1}}{4.00\text{ J}\left(\text{cm}^2\right)^{-1}\text{ min}^{-1}} = 6.25\text{ m}^2$

P5.31) Calculate ΔS° for the reaction $H_2(g) + Cl_2(g) \rightarrow 2HCl(g)$ at 650 K. Omit terms in the temperature-dependent heat capacities higher than $\dfrac{T^2}{\text{K}^2}$.

From Table 2.4,

$$C_P^\circ(\text{H}_2, g) = 22.66 + 4.38\times10^{-2}\frac{T}{\text{K}} - 1.0835\times10^{-4}\frac{T^2}{\text{K}^2}\text{ J K}^{-1}\text{mol}^{-1}$$

$$C_P^\circ(\text{Cl}_2, g) = 22.85 + 6.543\times10^{-2}\frac{T}{\text{K}} - 1.2517\times10^{-4}\frac{T^2}{\text{K}^2}\text{ J K}^{-1}\text{mol}^{-1}$$

$$C_P^\circ(\text{HCl}, g) = 29.81 - 4.12\times10^{-3}\frac{T}{\text{K}} + 6.2231\times10^{-6}\frac{T^2}{\text{K}^2}\text{ J K}^{-1}\text{mol}^{-1}$$

$$\Delta C_P^\circ = 2\left(29.81 - 4.12\times10^{-3}\frac{T}{\text{K}} + 6.2231\times10^{-6}\frac{T^2}{\text{K}^2}\text{ J K}^{-1}\text{mol}^{-1}\right)$$
$$-\left(22.66 + 4.38\times10^{-2}\frac{T}{\text{K}} - 1.0835\times10^{-4}\frac{T^2}{\text{K}^2}\text{ J K}^{-1}\text{mol}^{-1}\right)$$
$$-\left(22.85 + 6.543\times10^{-2}\frac{T}{\text{K}} - 1.2517\times10^{-4}\frac{T^2}{\text{K}^2}\text{ J K}^{-1}\text{mol}^{-1}\right)$$

$$\Delta C_P^\circ = 14.11 - 0.117\frac{T}{\text{K}} + 2.460\times10^{-4}\frac{T^2}{\text{K}^2}\text{ J K}^{-1}\text{mol}^{-1}$$

$$\Delta S^\circ = 2S^\circ_{298.15}(\text{HCl}, g) - S^\circ_{298.15}(\text{Cl}_2, g) - S^\circ_{298.15}(\text{H}_2, g)$$

$$= 2\times 186.9 \text{ J K}^{-1}\text{mol}^{-1} - 223.1 \text{ J K}^{-1}\text{mol}^{-1} - 130.7 \text{ J K}^{-1}\text{mol}^{-1}$$

$$= 22.0 \text{ J K}^{-1}\text{mol}^{-1}$$

$$\Delta S^\circ_T = \Delta S^\circ_{298.15} + \int_{298.15}^{T} \frac{\Delta C^\circ_p}{T'} dT'$$

$$= 20.0 \text{ J K}^{-1}\text{mol}^{-1} + \int_{298.15}^{650} \frac{\left(14.11 - 0.117\frac{T}{\text{K}} + 2.460\times 10^{-4}\frac{T^2}{\text{K}^2} \text{ J K}^{-1}\text{mol}^{-1}\right)}{\frac{T}{\text{K}}} d\frac{T}{\text{K}} \text{ J K}^{-1}\text{mol}^{-1}$$

$$= 20.0 \text{ J K}^{-1}\text{mol}^{-1} + 10.99 \text{ J K}^{-1}\text{mol}^{-1} - 41.33 \text{ J K}^{-1}\text{mol}^{-1} + 41.03 \text{ J K}^{-1}\text{mol}^{-1}$$

$$= 20.0 \text{ J K}^{-1}\text{mol}^{-1} + 10.69 \text{ J K}^{-1}\text{mol}^{-1} = 30.69 \text{ J K}^{-1}\text{mol}^{-1}$$

Chapter 6: Chemical Equilibrium

P6.1) Calculate the maximum nonexpansion work that can be gained from the combustion of benzene(l) and of $H_2(g)$ on a per gram and a per mole basis under standard conditions. Is it apparent from this calculation why fuel cells based on H_2 oxidation are under development for mobile applications?

$C_6H_6(l) + 15/2\ O_2(g) \rightarrow 6CO_2(g) + 3H_2O(l)$

$$w_{nonexpansion}^{max} = \Delta G_R^\circ = 3\Delta G_f^\circ\left(\mathrm{H_2O},l\right) + 6\Delta G_f^\circ\left(\mathrm{CO_2},g\right) - \frac{15}{2}\Delta G_f^\circ\left(\mathrm{O_2},g\right) - \Delta G_f^\circ\left(\mathrm{C_6H_6},l\right)$$

$$w_{nonexpansion}^{max} = 3\times\left(-237.1\ \mathrm{kJ\,mol^{-1}}\right) + 6\times\left(-394.4\ \mathrm{kJ\,mol^{-1}}\right) - \frac{15}{2}\times(0) - 124.5 - 124.5\ \mathrm{kJ\,mol^{-1}}$$

$$= -3202\ \mathrm{kJ\,mol^{-1}}$$

$$= -3202\ \mathrm{kJ\,mol^{-1}} \times \frac{1\ \mathrm{mol}}{78.18\ \mathrm{g}} = -40.96\ \mathrm{kJ\,g^{-1}}$$

$H_2(g) + 1/2\ O_2(g) \rightarrow H_2O(l)$

$$w_{nonexpansion}^{max} = \Delta G_R^\circ = \Delta G_f^\circ\left(\mathrm{H_2O},l\right) - \frac{1}{2}\Delta G_f^\circ\left(\mathrm{O_2},g\right) - \Delta G_f^\circ\left(\mathrm{H_2},g\right)$$

$$w_{nonexpansion}^{max} = -237.1\ \mathrm{kJ\,mol^{-1}} - 0 - 0$$

$$= -237.1\ \mathrm{kJ\,mol^{-1}}$$

$$= -237.1\ \mathrm{kJ\,mol^{-1}} \times \frac{1\ \mathrm{mol}}{2.016\ \mathrm{g}} = -117.6\ \mathrm{kJ\,g^{-1}}$$

On a per gram basis, nearly three times as much work can be extracted from the oxidation of hydrogen than benzene.

P6.3) Calculate ΔG for the isothermal expansion of 2.50 mol of an ideal gas at 350 K from an initial pressure of 10.5 bar to a final pressure of 0.500 bar.

$$dG = -SdT + VdP$$

At constant T, we consider the reversible process. Because G is a state function, any path between the same initial and final states will give the same result.

$$\Delta G = \int_{P_i}^{P_f} VdP = nRT\ln\frac{P_f}{P_i} = 2.50\ \mathrm{mol}\times 8.314\ \mathrm{J\,mol^{-1}K^{-1}}\times 350\ \mathrm{K}\times\ln\frac{0.500\ \mathrm{bar}}{10.5\ \mathrm{bar}} = -22.1\times 10^3\ \mathrm{J}$$

P6.5) The pressure dependence of G is quite different for gases and condensed phases. Calculate G_m(C, *solid, graphite*, 100 bar, 298.15 K) and G_m (He, g, 100 bar, 298.15 K) relative to their standard state values. By what factor is the change in G_m greater for He than for graphite?

For a solid or liquid,

$$\Delta G = \int_{P_i}^{P_f} VdP = V\left(P_f - P_i\right)$$

$$G_m(C,s,100\text{ bar}) = G_m(C,s,1\text{ bar}) + V_m\left(P_f - P_i\right) = G_m(C,s,1\text{ bar}) + \frac{M}{\rho}\left(P_f - P_i\right)$$

$$= 0 + \frac{12.011\times10^{-3}\text{kg}}{2250\text{ kg m}^{-3}}\times 99.0\times10^{5}\text{Pa} = 52.8\text{ J}$$

Treating He as an ideal gas,

$$G_m\left(He,g,100\text{ bar}\right) = G_m\left(He,g,1\text{ bar}\right) + \int_{P_i}^{P_f} VdP$$

$$= 0 + RT\ln\frac{P_f}{P_i} = 1\text{ mole}\times 8.314\text{ J mol}^{-1}\text{K}^{-1}\times 298.15\text{ K}\times\ln\frac{100\text{ bar}}{1\text{ bar}} = 11.4\times10^{3}\text{J}$$

This result is a factor of 216 greater than that for graphite.

P6.8) Calculate $\Delta A^{\circ}_{reaction}$ and $\Delta G^{\circ}_{reaction}$ for the reaction $CH_4(g) + 2O_2(g) \rightarrow CO_2(g) + 2H_2O(l)$ at 298 K from the combustion enthalpy of methane and the entropies of the reactants and products.

All reactants and products are treated as ideal gases

$$\Delta G^{\circ}_{combustion} = \Delta H^{\circ}_{combustion} - T\Delta S^{\circ}_{combustion}$$

$$\Delta S^{\circ}_{combustion} = S^{\circ}\left(\text{CO}_2,g\right) + 2S^{\circ}\left(\text{H}_2\text{O},l\right) - S^{\circ}\left(\text{CH}_4,g\right) - 2S^{\circ}\left(\text{O}_2,g\right)$$

$$= 213.8\text{ J mol}^{-1}\text{K}^{-1} + 2\times 70.0\text{ J mol}^{-1}\text{K}^{-1} - 186.3\text{ J mol}^{-1}\text{K}^{-1} - 2\times 205.2\text{ J mol}^{-1}\text{K}^{-1}$$

$$= -242.9\text{ J mol}^{-1}\text{K}^{-1}$$

$$\Delta G^{\circ}_{combustion} = -890.3\times10^{3}\text{J mol}^{-1} - 298.15\text{ K}\times\left(-242.7\text{ J mol}^{-1}\text{K}^{-1}\right) = -818.6\times10^{3}\text{J mol}^{-1}$$

$$\Delta A^{\circ}_{combustion} = \Delta U^{\circ}_{combustion} - T\Delta S^{\circ}_{combustion}$$

$$= \Delta H^{\circ}_{combustion} - \Delta\left(PV\right)_{combustion} - T\Delta S^{\circ}_{combustion}$$

$$= \Delta G^{\circ}_{combustion} + T\Delta S^{\circ}_{combustion} - \Delta\left(PV\right) - T\Delta S^{\circ}_{combustion}$$

$$= \Delta G^{\circ}_{combustion} - \Delta nRT$$

where Δn is the change in the number of moles of gas phase species in the reaction

$$= -818.6\times10^{3}\text{J mol}^{-1} - \left(-2\right)\times 8.314\text{J mol}^{-1}\text{K}^{-1}\times 298.15\text{ K}$$

$$\Delta A^{\circ}_{combustion} = -813.6\times10^{3}\text{J mol}^{-1}$$

P6.10) Consider the equilibrium $NO_2(g) \rightleftarrows NO(g) + 1/2O_2(g)$. 1 mol of $NO_2(g)$ is placed in a vessel and allowed to come to equilibrium at a total pressure of 1 atm. An analysis of the contents of the vessel gives the following results:

T	700 K	800 K
$\frac{P_{NO}}{P_{NO_2}}$	0.872	2.50

a) Calculate K_P at 700 and 800 K.

b) Calculate $\Delta G^{\circ}_{reaction}$ and $\Delta H^{\circ}_{reaction}$ for this reaction at 298.15 K. Assume that $\Delta H^{\circ}_{reaction}$ is independent of temperature.

a) $NO_2(g) \rightleftarrows NO(g) + \frac{1}{2}O_2(g)$

$$K_P = \frac{\left(P_{NO}/P^{\circ}\right)\left(P_{O_2}/P_{O}\right)}{P_{NO_2}/P^{\circ}}$$

At 700 K, $\frac{P_{NO}}{P_{NO_2}} = 0.872$ and $P_{O_2} = \frac{1}{2}P_{NO}$

$$P_{total} = P_{NO} + P_{NO_2} + P_{O_2} = 1 \text{ atm}$$

$$1 \text{ atm} = 0.872\,P_{NO_2} + P_{NO_2} + 0.436\,P_{NO_2}$$

$$1 \text{ atm} = 2.308\,P_{NO_2}$$

$$P_{NO_2} = 0.433 \text{ atm}$$

$$K_P = \frac{(0.872\times0.433)\times\sqrt{0.436\times0.433}}{0.433} = 0.379$$

At 800 K, $\frac{P_{NO}}{P_{NO_2}} = 2.50$ and $P_{O_2} = \frac{1}{2}P_{NO}$

$$P_{Total} = P_{NO} + P_{NO_2} + P_{O_2}$$

$$1 \text{ atm} = 2.50\,P_{NO_2} + P_{NO_2} + 1.25\,P_{NO_2} = 4.75\,P_{NO_2}$$

$$P_{NO_2} = 0.211 \text{ atm}$$

$$K_P = \frac{(2.50\times0.211)\times\sqrt{1.25\times0.211}}{0.211} = 1.284$$

b) Assuming that $\Delta H^{\circ}_{reaction}$ is independent of temperature,

$$\ln\frac{K_P(800\text{ K})}{K_P(700\text{ K})}=\frac{-\Delta H^\circ_{reaction}}{R}\left(\frac{1}{800\text{ K}}-\frac{1}{700\text{ K}}\right)$$

$$\Delta H^\circ_{reaction}=-\frac{R\times\ln\left(\dfrac{K_P(800\text{ K})}{K_P(700\text{ K})}\right)}{\left(\dfrac{1}{800\text{ K}}-\dfrac{1}{700\text{ K}}\right)}=56.8\times10^3\text{ J mol}^{-1}$$

$$\ln K_P(298.15\text{ K})=\ln K_P(700\text{ K})-\frac{\Delta H^\circ_{reaction}}{\text{R}}\times\left(\frac{1}{298.15\text{ K}}-\frac{1}{700\text{ K}}\right)=-14.12$$

$$\Delta G^\circ_{reaction}(298.15\text{ K})=-RT\ln K_P(298.15\text{ K})$$
$$=-8.314\text{ J mol}^{-1}\text{ K}^{-1}\times298.15\text{ K}\times(-14.12)$$
$$=35.0\times10^3\text{ J mol}^{-1}$$

P6.17) Calculate $\mu_{O_2}^{mixture}(298.15\text{ K, 1 bar})$ for oxygen in air, assuming that the mole

fraction of O_2 in air is 0.200.

We calculate the conventional molar Gibbs energy as described in Example Problem 6.3.

$$\mu^\circ_{O_2}(T)=-TS^\circ(\text{O}_2,g,298.15\text{ K})=-298.15\text{ K}\times205.2\text{ J K}^{-1}\text{mol}^{-1}=-61.2\text{ kJ mol}^{-1}$$

$$\mu_{O_2}^{mixture}(T,P)=\mu^\circ_{O_2}(T)+RT\ln\frac{P}{P^\circ}+RT\ln x_{O_2}$$
$$=-61.2\text{ kJ mol}^{-1}+RT\ln\frac{1\text{ bar}}{1\text{ bar}}+8.314\text{ J mol}^{-1}\text{K}^{-1}\times298.15\text{ K}\times\ln 0.200$$
$$=-65.2\times10^3\text{J mol}^{-1}$$

P6.22) Calculate K_P at 550 K for the reaction $N_2O_4(l)\rightarrow 2NO_2(g)$ assuming that $\Delta H^\circ_{reaction}$ is constant over the interval 298–600 K.

$$\Delta H^\circ_{reaction}=2\Delta H^\circ_f(\text{NO}_2,g)-\Delta H^\circ_f(\text{N}_2\text{O}_4,l)$$
$$=2\times33.2\times10^3\text{J mol}^{-3}+19.5\times10^3\text{J mol}^{-3}=85.9\times10^3\text{J mol}^{-3}$$
$$\Delta G^\circ_{reaction}=2\Delta G^\circ_f(\text{NO}_2,g)-\Delta G^\circ_f(\text{N}_2\text{O}_4,l)$$
$$=2\times51.3\times10^3\text{J mol}^{-3}-97.5\times10^3\text{J mol}^{-3}=5.1\times10^3\text{J mol}^{-3}$$

$$\ln K_P(T_f) = -\frac{\Delta G^\circ_{reaction}(298.15\text{ K})}{R\times 298.15\text{ K}} - \frac{\Delta H^\circ_{reaction}}{R}\left(\frac{1}{T_f} - \frac{1}{298.15\text{ K}}\right)$$

$$\ln K_P(550\text{ K}) = -\frac{5.1\times10^3\text{J mol}^{-1}}{8.314\text{ J K}^{-1}\text{mol}^{-1}\times 298.15\text{ K}} - \frac{85.9\times10^3\text{J mol}^{-1}}{8.314\text{ J K}^{-1}\text{mol}^{-1}}\times\left(\frac{1}{550\text{ K}} - \frac{1}{298.15\text{ K}}\right)$$

$$\ln K_P(550\text{ K}) = 13.81$$

$$K_P(550\text{ K}) = 9.95\times10^5$$

P6.24) Calculate the degree of dissociation of N_2O_4 in the reaction $N_2O_4(g) \rightarrow 2NO_2(g)$ at 250 K and a total pressure of 0.500 bar. Do you expect the degree of dissociation to increase or decrease as the temperature is increased to 550 K? Assume that $\Delta H^\circ_{reaction}$ is independent of temperature.

We set up the following table

	$N_2O_4(g)$	$\rightarrow$	$2NO_2(g)$
Initial number of moles	n_0		0
Moles present at equilibrium	$n_0 - \xi$		2ξ
Mole fraction present at equilibrium	$\frac{n_0 - \xi}{n_0 + \xi}$		$\frac{2\xi}{n_0 + \xi}$
Partial pressure at equilibrium, $P_i = x_i P$	$\left(\frac{n_0 - \xi}{n_0 + \xi}\right)P$		$\left(\frac{2\xi}{n_0 + \xi}\right)P$

We next express K_P in terms of n_0, ξ, and P.

$$K_P(T) = \frac{\left(\frac{P^{eq}_{Cl}}{P^\circ}\right)^2}{\left(\frac{P^{eq}_{Cl_2}}{P^\circ}\right)} = \frac{\left[\left(\frac{2\xi}{n_0+\xi}\right)\frac{P}{P^\circ}\right]^2}{\left(\frac{n_0-\xi}{n_0+\xi}\right)\frac{P}{P^\circ}} = \frac{4\xi^2}{(n_0+\xi)(n_0-\xi)}\frac{P}{P^\circ} = \frac{4\xi^2}{(n_0)^2 - \xi^2}\frac{P}{P^\circ}$$

We convert this expression for K_P to one in terms of α.

$$K_P(T) = \frac{4\xi^2}{(n_0)^2 - \xi^2}\frac{P}{P^\circ} = \frac{4\alpha^2}{1-\alpha^2}\frac{P}{P^\circ}$$

$$\left(K_P(T) + 4\frac{P}{P^\circ}\right)\alpha^2 = K_P(T)$$

$$\alpha = \sqrt{\frac{K_P(T)}{K_P(T) + 4\frac{P}{P^\circ}}}$$

$$\Delta G^\circ_{reaction} = 2\Delta G^\circ_f(\mathrm{NO_2}, g) - \Delta G^\circ_f(\mathrm{N_2O_4}, g)$$
$$= 2\times 51.3\times 10^3 \,\mathrm{J\,mol^{-3}} - 99.8\times 10^3 \,\mathrm{J\,mol^{-3}} = 2.8\times 10^3 \,\mathrm{J\,mol^{-3}}$$

$$\Delta H^\circ_{reaction} = 2\Delta H^\circ_f(\mathrm{NO_2}, g) - \Delta H^\circ_f(\mathrm{N_2O_4}, g)$$
$$= 2\times 33.2\times 10^3 \,\mathrm{J\,mol^{-3}} - 11.1\times 10^3 \,\mathrm{J\,mol^{-3}} = 55.3\times 10^3 \,\mathrm{J\,mol^{-3}}$$

$$\ln K_P(T_f) = -\frac{\Delta G^\circ_{reaction}(298.15\ \mathrm{K})}{R\times 298.15\ \mathrm{K}} - \frac{\Delta H^\circ_{reaction}}{R}\left(\frac{1}{T_f} - \frac{1}{298.15\ \mathrm{K}}\right)$$

$$\ln K_P(250\ \mathrm{K}) = -\frac{2.8\times 10^3 \,\mathrm{J\,mol^{-1}}}{8.314\ \mathrm{J\,K^{-1}mol^{-1}}\times 298.15\ \mathrm{K}} - \frac{55.3\times 10^3 \,\mathrm{J\,mol^{-1}}}{8.314\ \mathrm{J\,K^{-1}mol^{-1}}}\times\left(\frac{1}{250\ \mathrm{K}} - \frac{1}{298.15\ \mathrm{K}}\right)$$

$$\ln K_P(250\ \mathrm{K}) = -5.426$$

$$K_P(250\ \mathrm{K}) = 4.40\times 10^{-3}$$

$$\alpha = \sqrt{\frac{4.40\times 10^{-3}}{4.40\times 10^{-3} + 4\times 0.500}} = 4.68\times 10^{-2}$$

Because $\Delta H^\circ_{reaction} > 0$, *a* increases as *T* increases.

P6.26) A sample containing 2.00 moles of N_2 and 6.00 mol of H_2 are placed in a reaction vessel and brought to equilibrium at 20.0 bar and 750 K in the reaction 1/2 $N_2(g)$ + 3/2 $H_2(g) \rightarrow NH_3(g)$.

a) Calculate K_P at this temperature.

b) Set up an equation relating K_P and the extent of reaction as in Example Problem 6.9.

c) Using a numerical equation solver, calculate the number of moles of each species present at equilibrium.

	1/2 $N_2(g)$ +	3/2 $H_2(g)$ →	$NH_3(g)$
Initial number of moles	2.00	6.00	0
Moles present at equilibrium	$2.00 - \xi$	$6.00 - 3\xi$	2ξ

Mole fraction present at equilibrium	$\frac{2.00-\xi}{8-2\xi}$	$\frac{6.00-3\xi}{8-2\xi}$	$\frac{2\xi}{8-2\xi}$
Partial pressure at equilibrium, $P_i = x_i P$	$\left(\frac{2.00-\xi}{8-2\xi}\right)P$	$\left(\frac{6.00-3\xi}{8-2\xi}\right)P$	$\left(\frac{2\xi}{8-2\xi}\right)P$

We next express K_P in terms of n_0, ξ, and P.

$$K_P(T) = \frac{\left(\frac{P^{eq}_{NH_3}}{P^\circ}\right)}{\left(\frac{P^{eq}_{N_2}}{P^\circ}\right)^{\frac{1}{2}}\left(\frac{P^{eq}_{H_2}}{P^\circ}\right)^{\frac{3}{2}}} = \frac{\left(\frac{2\xi}{8-2\xi}\right)\frac{P}{P^\circ}}{\left(\left(\frac{2.00-\xi}{8-2\xi}\right)\frac{P}{P^\circ}\right)^{\frac{1}{2}}\left(\left(\frac{6.00-3\xi}{8-2\xi}\right)\frac{P}{P^\circ}\right)^{\frac{3}{2}}}$$

$$\Delta H^\circ_{reaction} = \Delta H^\circ_f(\mathrm{NH_3}, g) = -45.9\times10^3\,\mathrm{J\,mol^{-1}}$$

$$\Delta G^\circ_{reaction} = \Delta G^\circ_f(\mathrm{NH_3}, g) = -16.5\times10^3\,\mathrm{J\,mol^{-1}}$$

$$\ln K_P(T_f) = -\frac{\Delta G^\circ_{reaction}(298.15\text{ K})}{R\times298.15\text{ K}} - \frac{\Delta H^\circ_{reaction}}{R}\left(\frac{1}{T_f} - \frac{1}{298.15\text{ K}}\right)$$

$$\ln K_P(750K) = \frac{16.5\times10^3\,\mathrm{J\,mol^{-1}}}{8.314\,\mathrm{J\,mol^{-1}K^{-1}}\times298.15\text{ K}} + \frac{45.9\times10^3\,\mathrm{J\,mol^{-1}}}{8.314\,\mathrm{J\,mol^{-1}K^{-1}}}\times\left(\frac{1}{750\text{ K}} - \frac{1}{298.15\text{ K}}\right)$$

$$= -4.4994$$

$$K_P(750\text{ K}) = 1.11\times10^{-2}$$

The following equation can be solved numerically using a program such as Mathematica.

$$K_P(T) = \frac{\left(\frac{2\xi}{8-2\xi}\right)20}{\left(\left(\frac{2.00-\xi}{8-2\xi}\right)20\right)^{\frac{1}{2}}\left(\left(\frac{6.00-3\xi}{8-2\xi}\right)20\right)^{\frac{3}{2}}} = 1.11\times10^{-2}$$

The physically meaningful root of the cubic equation is $\xi = 0.241$. Therefore, there are 1.76 moles of $N_2(g)$, 5.28 moles of $H_2(g)$, and 0.48 moles of $NH_3(g)$ at equilibrium.

P6.27) Consider the equilibrium in the reaction $3O_2(g) \rightleftarrows 2O_3(g)$, with $\Delta H^\circ_{reaction} = 285.4 \times 10^3$ J mol^{-1} at 298 K. Assume that $\Delta H^\circ_{reaction}$ is independent of temperature.

a) Without doing a calculation, predict whether the equilibrium position will shift toward reactants or products as the pressure is increased.

b) Without doing a calculation, predict whether the equilibrium position will shift toward reactants or products as the temperature is increased.

c) Calculate K_P at 550 K.

d) Calculate K_x at 550 K and 0.500 bar.

a) Without doing a calculation, predict if the equilibrium position will shift towards reactants or products as the pressure is increased.

The number of moles of products is fewer than the number of moles of reactants. Therefore, the equilibrium position will shift towards products as the pressure is increased.

b) Without doing a calculation, predict if the equilibrium position will shift towards reactants or products as the temperature is increased.

Because $\Delta H^\circ_{reaction} > 0$, the equilibrium position will shift towards products as the temperature is increased.

c) Calculate K_P at 550 K.

$$\Delta G^\circ_{reaction} = 2\Delta G^\circ_f(\text{O}_3, g) = 2\times163.2\times10^3\,\text{J mol}^{-1}$$

$$\Delta H^\circ_{reaction} = 2\Delta H^\circ_f(\text{O}_3, g) = 2\times142.7\times10^3\,\text{J mol}^{-1}$$

$$\ln K_P(T_f) = -\frac{\Delta G^\circ_{reaction}(298.15\text{ K})}{R\times298.15\text{ K}} - \frac{\Delta H^\circ_{reaction}}{R}\left(\frac{1}{T_f} - \frac{1}{298.15\text{ K}}\right)$$

$$\ln K_P(550\text{ K}) = -\frac{2\times163.2\times10^3\text{J mol}^{-1}}{8.314\text{ J K}^{-1}\text{mol}^{-1}\times298.15\text{ K}} - \frac{2\times142.7\times10^3\text{J mol}^{-1}}{8.314\text{ J K}^{-1}\text{mol}^{-1}}\times\left(\frac{1}{550\text{ K}} - \frac{1}{298.15\text{ K}}\right)$$

$$\ln K_P(550\text{ K}) = -78.954$$

$$K_P(550\text{ K}) = 5.13\times10^{-35}$$

d) Calculate K_X at 550 K and 0.500 bar.

$$K_X = K_P\left(\frac{P}{P^\circ}\right)^{-\Delta\nu} = 5.13\times10^{-35}\times\left(\frac{0.500\text{ bar}}{1\text{ bar}}\right)^{-1} = 1.03\times10^{-34}$$

Chapter 7: The Properties of Real Gases

P7.2) Calculate the density of $O_2(g)$ at 375 K and 385 bar using the ideal gas and the van der Waals equations of state. Use a numerical equation solver to solve the van der Waals equation for V_m or use an iterative approach starting with V_m equal to the ideal gas result. Based on your result, does the attractive or repulsive contribution to the interaction potential dominate under these conditions?

$$V_m = \frac{RT}{P} = \frac{8.314\times10^{-2}\,\text{L bar K}^{-1}\text{mol}^{-1}\times 375\text{ K}}{385\text{ bar}} = 8.10\times10^{-2}\text{L}$$

$$P_{vdW} = \frac{RT}{V_m - b} - \frac{a}{V_m^2} = \frac{8.314\times10^{-2}\,\text{L bar K}^{-1}\text{mol}^{-1}\times 370\text{ K}}{V_m - 0.0319\text{ L mol}^{-1}} - \frac{1.382\text{ L}^2\text{ bar mol}^{-2}}{(V_m)^2}$$

The three solutions to this equation are

$V_m = (0.0131 \pm 0.0339i)\,\text{L mol}^{-1}$ and $V_m = 0.0867\text{ L mol}^{-1}$

Only the real solution is of significance.

$$\rho_{idealgas} = \frac{M}{V_m} = \frac{32.0\text{ g mol}^{-1}}{0.0810\text{ L mol}^{-1}} = 395\text{ g L}^{-1}$$

$$\rho_{vdW} = \frac{M}{V_m} = \frac{32.0\text{ g mol}^{-1}}{0.0867\text{ L mol}^{-1}} = 369\text{ g L}^{-1}$$

Because the van der Waals density is less than the ideal gas density, the repulsive part of the potential dominates.

P7.5) Calculate the van der Waals parameters of methane from the values of the critical constants.

$$b = \frac{RT_c}{8P_c} = \frac{8.314\times10^{-2}\text{dm}^3\text{bar K}^{-1}\text{mol}^{-1}\times 190.56\text{ K}}{8\times 45.99\text{ bar}} = 0.0431\text{ dm}^3\text{mol}^{-1}$$

$$a = \frac{27R^2T_c^2}{64P_c} = \frac{27\times\left(8.314\times10^{-2}\text{dm}^3\text{bar K}^{-1}\text{mol}^{-1}\right)^2\times(190.56\text{ K})^2}{64\times 45.99\text{ bar}} = 2.303\text{ dm}^6\text{bar mol}^{-2}$$

P7.7) Use the law of corresponding states and Figure 7.8 to estimate the molar volume of methane at $T = 285$ K and $P = 180$ bar.

$$T_r = \frac{285\text{ K}}{190.56\text{ K}} = 1.5 \quad P_r = \frac{180\text{ bar}}{45.99\text{ bar}} = 3.91. \text{ Therefore, } z \approx 0.8.$$

$$\frac{PV_m}{RT} = 0.8; \quad V_m = 0.8\frac{RT}{P} = 0.8\times\frac{8.314\times10^{-2}\text{dm}^3\text{bar K}^{-1}\text{mol}^{-1}\times 285\text{ K}}{180\text{ bar}}$$

$V_m = 0.105\text{ L mol}^{-1}$

P7.9) Assume that the equation of state for a gas can be written in the form $P(V_m - b(T)) = RT$. Derive an expression for $\beta = \frac{1}{V}\left(\frac{\partial V}{\partial T}\right)_P$ and $\kappa = -\frac{1}{V}\left(\frac{\partial V}{\partial P}\right)_T$ for such a gas in terms of $b(T), \frac{db(T)}{dT}, P,$ and V_m.

$$P\left(\frac{V}{n} - b(T)\right) = RT; \quad \frac{V}{n} = \frac{RT}{P} + b(T)$$

$$V = nb(T) + \frac{nRT}{P}$$

$$\beta = \frac{1}{V}\left(\frac{\partial V}{\partial T}\right)_P = \frac{1}{V}\left(\frac{ndb(T)}{dT} + \frac{nR}{P}\right) = \frac{1}{V_m}\left(\frac{db(T)}{dT} + \frac{R}{P}\right)$$

$$\kappa = -\frac{1}{V}\left(\frac{\partial V}{\partial P}\right)_T = -\frac{1}{V}\left(-\frac{nRT}{P^2}\right) = \frac{RT}{V_m P^2}$$

P7.12) For a van der Waals gas, $z = \frac{V_m}{V_m - b} - \frac{a}{RTV_m}$. Expand the first term of this expression in a Taylor series in the limit $V_m >> b$ to obtain $z \approx 1 + \left(b - \frac{a}{RT}\right)\frac{1}{V_m}$.

$f(x) = f(0) + \left(\frac{df(x)}{dx}\right)_{x=0} x + \ldots$ In this case, $f(x) = \frac{1}{1 - \frac{b}{V_m}}$ and $x = \frac{b}{V_m}$

$$z = \frac{V_m}{V_m - b} - \frac{a}{RTV_m} = \frac{1}{1 - \frac{b}{V_m}} - \frac{a}{RTV_m}$$

$$\frac{1}{1 - \frac{b}{V_m}} \approx \left(1 \Big/ \left[1 - \frac{b}{V_m}\right]\right)_{\frac{b}{V_m}=1} + \left[\frac{d\left(1 \Big/ \left[1 - \frac{b}{V_m}\right]\right)}{d\left(\frac{b}{V_m}\right)}\right]_{\frac{b}{V_m}=1} \frac{b}{V_m} = 1 + \frac{b}{V_m}$$

$$z \approx 1 + \frac{b}{V_m} - \frac{a}{RTV_m} = 1 + \frac{1}{V_m}\left(b - \frac{a}{RT}\right)$$

P7.16) Another equation of state is the Bertholet equation, $V_m = \frac{RT}{P} + b - \frac{a}{RT^2}$. Derive expressions for $\beta = \frac{1}{V}\left(\frac{\partial V}{\partial T}\right)_P$ and $\kappa = -\frac{1}{V}\left(\frac{\partial V}{\partial P}\right)_T$ from the Bertholet equation in terms of V, T, and P.

$$V = \frac{nRT}{P} + nb - \frac{na}{RT^2}$$

$$\beta = \frac{1}{V}\left(\frac{\partial V}{\partial T}\right)_P = \frac{1}{V}\left(\frac{nR}{P} + \frac{2na}{RT^3}\right) = \frac{1}{V_m}\left(\frac{R}{P} + \frac{2a}{RT^3}\right)$$

$$\kappa = -\frac{1}{V}\left(\frac{\partial V}{\partial P}\right)_T = -\frac{1}{V}\left(-\frac{nRT}{P^2}\right) = \frac{nRT}{P^2V} = \frac{RT}{P^2V_m}$$

P7.18) The experimentally determined density of H_2O at 1200 bar and 800 K is 537 g L^{-1}. Calculate z and V_m from this information. Compare this result with what you would have estimated from Figure 7.8. What is the relative error in using Figure 7.8 for this case?

$$V_m = \frac{M}{\rho} = \frac{18.02 \text{ g mol}^{-1}}{537 \text{ g L}^{-1}} = 3.34 \times 10^{-2} \text{L mol}^{-1}$$

$$z = \frac{PV_m}{RT} = \frac{1200 \text{ bar} \times 3.34 \times 10^{-2} \text{L mol}^{-1}}{8.314 \times 10^{-2} \text{L bar mol}^{-1}\text{K}^{-1} \times 800 \text{ K}} = 0.602$$

Because $P_r = \frac{1200 \text{ bar}}{220.64 \text{ bar}} = 5.44$ and $T_r = \frac{800 \text{ K}}{647.14 \text{ K}} = 1.24$, Figure 7.7 predicts $z = 0.75$.
The relative error in z is 25%.

P7.20) At what temperature does the slope of the z versus P curve as $P \rightarrow 0$ have its maximum value for a van der Waals gas? What is the value of the maximum slope?

$$\left(\frac{\partial Z}{\partial P}\right)_{T,P\to 0} = \frac{1}{RT}\left(b - \frac{a}{RT}\right) \text{ for a van der Waals gas}$$

$$\left(\frac{\partial}{\partial T}\left(\frac{\partial Z}{\partial P}\right)_T\right)_{P\to 0} = -\frac{1}{RT^2}\left(b - \frac{a}{RT}\right) + \frac{1}{RT^3} = -\frac{1}{RT^2}\left(b - \frac{2a}{RT}\right)$$

Setting this derivative equal to zero gives

$$b - \frac{2a}{RT_{\max}} = 0 \quad T_{\max} = \frac{2a}{Rb}$$

The maximum slope is $\frac{1}{RT_{\max}}\left(b - \frac{a}{RT_{\max}}\right) = \frac{b}{2a}\left[b - a\left(\frac{b}{2a}\right)\right] = \frac{b^2}{4a}$

Chapter 8: The Relative Stability of Solids, Liquids, and Gases

P8.3) Within what range can you restrict the values of *P* and *T* if the following information is known about CO_2? Use Figure 8.10 to answer this question.
a) As the temperature is increased, the solid is first converted to the liquid and subsequently to the gaseous state.

The temperature and pressure are greater than the values for the triple point, –56.6°C and 5.11 atm.

b) As the pressure on a cylinder containing pure CO_2 is increased from 65 to 80 atm, no interface delineating liquid and gaseous phases is observed.

The temperature is greater than the critical temperature, 31.0°C.

c) Solid, liquid, and gas phases coexist at equilibrium.

The system is at the triple point, –56.6°C and 5.11 atm.

d) An increase in pressure from 10 to 50 atm converts the liquid to the solid.

The temperature is slightly greater than the triple point value of –56.6°C.

e) An increase in temperature from –80° to 20°C converts a solid to a gas with no intermediate liquid phase.

The pressure is below the triple point pressure value of 5.11 atm.

P8.5) The vapor pressure of liquid SO_2 is 2232 Pa at 201 K, and $\Delta H_{vaporization} = 24.94$ kJ mol^{-1}. Calculate the normal and standard boiling points. Does your result for the normal boiling point agree with that in Table 8.2 set? If not, suggest a possible cause.

$$\ln\frac{P_f}{P_i} = -\frac{\Delta H_m^{vaporization}}{R}\left(\frac{1}{T_f}-\frac{1}{T_i}\right)$$

$$T_f = \frac{\Delta H_m^{vaporization}}{R\left(\dfrac{\Delta H_m^{vaporization}}{RT_i} - \ln\dfrac{P_f}{P_i}\right)}$$

At the normal boiling point, P = 101325 Pa.

$$T_{b,normal} = \frac{24.94\times10^3\,\mathrm{J\,mol^{-1}}}{8.314\,\mathrm{J\,mol^{-1}K^{-1}}\times\left(\dfrac{24.94\times10^3\,\mathrm{J\,mol^{-1}}}{8.314\,\mathrm{J\,mol^{-1}K^{-1}}\times 201\mathrm{K}} - \ln\dfrac{101325}{2232}\right)} = 271.8\ \mathrm{K}$$

At the standard boiling point, $P = 10^5$Pa.

$$T_{b,standard} = \frac{24.94\,\mathrm{x}\,10^3\,\mathrm{J\,mol^{-1}}}{8.314\,\mathrm{J\,mol^{-1}K^{-1}}\times\left(\dfrac{24.94\times10^3\,\mathrm{J\,mol^{-1}}}{8.314\,\mathrm{J\,mol^{-1}K^{-1}}\times 201\mathrm{K}} - \ln\dfrac{100000}{2232}\right)} = 269.6\ \mathrm{K}$$

The result for the normal boiling point is ~8 K higher than the value tabulated in Table 8.2. The most probable reason for this difference is that the calculation above has assumed that $\Delta H_{vaporization}$ is independent of T.

P8.7) Use the values for ΔG_f (ethanol, l) and ΔG_f (ethanol, g) in Appendix A to calculate the vapor pressure of ethanol at 298.15 K.

For the transformation C_2H_5OH (l) → C_2H_5OH (g)

$$\ln K_P = \frac{\ln P_{C_2H_5OH(g)}}{P^\circ} = -\frac{\Delta G_f^\circ(\mathrm{C_2H_5OH},g) - \Delta G_f^\circ(\mathrm{C_2H_5OH},l)}{RT}$$

$$= -\frac{-167.9\times10^3\,\mathrm{J\,mol^{-1}} + 174.8\times10^3\,\mathrm{J\,mol^{-1}}}{8.314\,\mathrm{J\,mol^{-1}K^{-1}}\times 298\ \mathrm{K}} = 2.785$$

$$K_P = \frac{P_{C_2H_5OH(g)}}{1\ \mathrm{bar}} = 0.0617\ \mathrm{bar} = 6.17\times10^3\mathrm{Pa}$$

P8.14) It has been suggested that the surface melting of ice plays a role in enabling speed skaters to achieve peak performance. Carry out the following calculation to test this hypothesis. At 1 atm pressure, ice melts at 273.15 K, ΔH_{fusion} = 6010 J mol^{-1}, the density of ice is 920 kg m^{-3}, and the density of liquid water is 997 kg m^{-3}.
a) What pressure is required to lower the melting temperature by 5.0°C?
b) Assume that the width of the skate in contact with the ice has been reduced by sharpening to 25 × 10^{-3} cm, and that the length of the contact area is 15 cm. If a skater of mass 85 kg is balanced on one skate, what pressure is exerted at the interface of the skate and the ice?
c) What is the melting point of ice under this pressure?

d) If the temperature of the ice is –5.0°C, do you expect melting of the ice at the ice–skate interface to occur?

a) What pressure is required to lower the melting temperature by 5.0°C?

$$\left(\frac{dP}{dT}\right)_{fusion} = \frac{\Delta S_m^{fusion}}{\Delta V_m^{fusion}} \approx \frac{\Delta S_m^{fusion}}{\frac{M}{\rho_{H_2O,l}} - \frac{M}{\rho_{H_2O,l}}} = \frac{22.0\ \text{J mol}^{-1}\text{K}^{-1}}{\frac{18.02\times10^{-3}\text{kg}}{998\ \text{kg m}^{-3}} - \frac{18.02\times10^{-3}\text{kg}}{920\ \text{kg m}^{-3}}}$$

$$= -1.44\times10^{7}\text{Pa K}^{-1} = -144\ \text{bar K}^{-1}$$

The pressure must be increased by 720 bar to lower the melting point by 5.0°C.

b) Assume that the width of the skate in contact with the ice is 25×10^{-3} cm, and that the length of the contact area is 15 cm. If a skater of mass 85 kg is balanced on one skate, what pressure is exerted at the interface of the skate and the ice?

$$P = \frac{F}{A} = \frac{85\ \text{kg}\times9.81\ \text{m s}^{-2}}{15\times10^{-2}\text{m}\times25\times10^{-5}\text{m}} = 2.2\times10^{7}\text{Pa} = 2.2\times10^{2}\text{bar}$$

c) What is the melting point of ice under this pressure?

$$\Delta T = \left(\frac{dT}{dP}\right)_{fusion} \Delta P = \frac{1^\circ\text{C}}{144\ \text{bar}} \times 2.20\times10^{2}\text{bar} = 1.5^\circ\text{C}\,;\ T_m = -1.5^\circ\text{C}$$

d) If the temperature of the ice is –5.0°C, do you expect melting of the ice at the ice-skate interface to occur?

No, because the lowering of the melting temperature is less than the temperature of the ice.

P8.15) Solid iodine, $I_2(s)$, at 25°C has an enthalpy of sublimation of 56.30 kJ mol^{-1}. The $C_{P,m}$ of the vapor and solid phases at that temperature are 36.9 and 54.4 J K^{-1} mol^{-1}, respectively. The sublimation pressure at 25°C is 0.30844 Torr. Calculate the sublimation pressure of the solid at the melting point (113.6°C) assuming
a) that the enthalpy of sublimation and the heat capacities do not change with temperature.
b) that the enthalpy of sublimation at temperature T can be calculated from the equation $\Delta H^\circ_{sublimation}(T) = \Delta H^\circ_{sublimation}(T_0) + \Delta C_P(T - T_0)$.

a) If the enthalpy of sublimation is constant

$$\ln\frac{P_2}{P_1} = -\frac{\Delta H^\circ_{sublimation}}{R}\left(\frac{1}{T_2}-\frac{1}{T_1}\right)$$

$$\ln P_2 = \ln 0.30844 - \frac{56.30\times10^3 \text{ J mol}^{-1}}{8.314 \text{ J mol}^{-1}\text{ K}^{-1}}\times\left(\frac{1}{386.8 \text{ K}}-\frac{1}{298.15 \text{ K}}\right)$$

$$P_2 = 56.22 \text{ Torr}$$

b) If the enthalpy of sublimation is given by $\Delta H^\circ_{sublimation}(T) = \Delta H^\circ_{sublimation}(T_0)+\Delta C_P(T-T_0)$

$$\int\frac{dP}{P} = \int_{T_0}^{T}\frac{\Delta H^\circ_{sublimation}}{RT^2}dT = \int_{T_0}^{T}\frac{\Delta H^\circ_{sublimation}(T_0)+\Delta C_P(T-T_0)}{RT^2}dT$$

$$\ln\frac{P_2}{P_1} = -\frac{\Delta H^\circ_{sublimation}(T_0)}{R}\left(\frac{1}{T}-\frac{1}{T_0}\right)+\frac{\Delta C_P T_0}{R}\left[\frac{1}{T}-\frac{1}{T_0}\right]+\frac{\Delta C_P}{R}\ln\frac{T}{T_0}$$

$$\ln P_2 = \ln 0.30844 - \frac{56.30\times10^3 \text{ J mol}^{-1}}{8.314 \text{ J K}^{-1}\text{ mol}^{-1}}\times\left(\frac{1}{386.8 \text{ K}}-\frac{1}{298.15 \text{ K}}\right)$$
$$+\frac{(36.9-54.4)\text{J K}^{-1}\text{ mol}^{-1}\times298.15 \text{ K}}{8.314 \text{ J K}^{-1}\text{ mol}^{-1}}\times\left(\frac{1}{386.8 \text{ K}}-\frac{1}{298.15 \text{ K}}\right)$$
$$+\frac{(36.9-54.4)\text{J K}^{-1}\text{ mol}^{-1}}{8.314 \text{ J K}^{-1}\text{ mol}^{-1}}\times\ln\frac{386.8 \text{ K}}{298.15 \text{ K}}$$

$$\ln P_2 = 3.964$$

$$P_2 = 52.65 \text{ Torr}$$

P8.17) Consider the transition between two forms of solid tin, $\text{Sn}(s,\text{gray}) \rightleftharpoons \text{Sn}(s,\text{white})$. The two phases are in equilibrium at 1 bar and 18°C. The densities for gray and white tin are 5750 and 7280 kg m^{-3}, respectively, and $\Delta S_{transition}$ = 8.8 J K^{-1} mol^{-1}. Calculate the temperature at which the two phases are in equilibrium at 200 bar.

In going from 1 atm, 18°C to 200 atm, and the unknown temperature T

$$\Delta G^{gray} = V_m^{gray}\Delta P - S^{gray}\Delta T$$

$$\Delta G^{white} = V_m^{white}\Delta P - S^{white}\Delta T$$

At equilibrium

$$\Delta G^{gray} - \Delta G^{white} = 0 = \left(V_m^{gray} - V_m^{white}\right)\Delta P - \left(S^{gray} - S^{white}\right)\Delta T$$

$$\Delta T = \frac{\left(V_m^{gray} - V_m^{white}\right)\Delta P}{\left(S^{gray} - S^{white}\right)} = \frac{M_{Sn}\left(\frac{1}{\rho_{gray}} - \frac{1}{\rho_{white}}\right)\Delta P}{\Delta S_{transition}}$$

$$= \frac{118.71\times10^{-3}\text{kg mol}^{-1}\times\left(\frac{1}{5750\text{ kg m}^{-3}} - \frac{1}{7280\text{ kg m}^{-3}}\right)\times199\text{ x }10^5\text{Pa}}{-8.8\text{ J K}^{-1}\text{mol}^{-1}} = -9.8^{\circ}\text{C}$$

$$T_f = 8.2^{\circ}\text{C}$$

P8.19) The vapor pressure of methanol (l) is given by $\ln\left(\frac{P}{\text{Pa}}\right) = 23.593 - \frac{3.6791\times10^3}{\frac{T}{\text{K}} - 31.317}$.

a) Calculate the standard boiling temperature.
b) Calculate $\Delta H_{vaporization}$ at 298 K and at the standard boiling temperature.

a)

$$\ln\left(\frac{P}{\text{Pa}}\right) = 23.593 - \frac{3.6791\times10^3}{\frac{T_b}{\text{K}} - 31.317} = \ln 10^5 = 11.5129$$

$$3.6791\times10^3 = (23.593 - 11.5129)\left(\frac{T_b}{\text{K}} - 31.317\right)$$

$$\frac{T_b}{\text{K}} = \frac{3.6791\times10^3}{(23.593 - 11.5129)} + 31.317 = 335.9$$

b)

$$\Delta H_{vaporization} = RT^2\frac{d\ln P}{dT} = \frac{8.314\text{ J mol}^1\text{ K}^{-1}\times3679.1\times(298\text{ K})^2}{(T - 31.37)^2}$$

$$= 38.19\text{ kJ mol}^{-1}\text{ at 298 K and }37.20\text{ kJ mol}^{-1}\text{ at 335.9 K}$$

P8.22) The densities of a given solid and liquid of molecular weight 122.5 at its normal melting temperature of 427.15 K are 1075 and 1012 kg m^{-3}, respectively. If the pressure is increased to 120 bar, the melting temperature increases to 429.35 K. Calculate $\Delta H^{\circ}_{fusion}$ and $\Delta S^{\circ}_{fusion}$ for this substance.

$$\frac{\Delta P}{\Delta T} \approx \frac{\Delta S}{\Delta V}; \quad \Delta S \approx \frac{\Delta P}{\Delta T}\Delta V$$

$$\Delta S_{fusion} = \frac{\Delta P}{\Delta T} M\left(\frac{1}{\rho_{liquid}} - \frac{1}{\rho_{solid}}\right)$$

$$\Delta S_{fusion} = \frac{119\times10^5 \text{Pa}}{429.35\text{ K} - 427.15\text{ K}} \times 122.5\times10^{-3}\text{kg mol}^{-1} \times \left(\frac{1}{1012\text{ kg m}^{-3}} - \frac{1}{1075\text{ kg m}^{-3}}\right)$$

$$= 38.4\text{ J K}^{-1}\text{mol}^{-1}$$

$$\Delta H_{fusion} = T_{fusion}\Delta S_{fusion} = 427.15\text{ K}\times 38.4\text{ J K}^{-1}\text{mol}^{-1} = 16.4\times10^3\text{J mol}^{-1} \text{ at 1 bar}$$

P8.24) The variation of the vapor pressure of the liquid and solid forms of a pure substance near the triple point are given by $\ln\frac{P_{solid}}{\text{Pa}} = -8750\frac{\text{K}}{T} + 31.143$ and $\ln\frac{P_{liquid}}{\text{Pa}} = -4053\frac{\text{K}}{T} + 21.10$. Calculate the temperature and pressure at the triple point.

At the triple point, $P_{solid} = P_{liquid}$.

$$-8750\frac{\text{K}}{T} + 31.143 = -4053\frac{\text{K}}{T} + 21.10$$

$$31.143 - 21.10 = (8750 - 4053)\frac{\text{K}}{T}$$

$$T = \frac{(8750 - 4053)}{31.143 - 21.10} = 467.7\text{ K}$$

$$\ln\frac{P_{tp}}{\text{Pa}} = \frac{-8750}{467.7} + 31.143 = 12.4344$$

$$\frac{P_{tp}}{\text{Pa}} = 2.513\times10^5$$

P8.30) Autoclaves that are used to sterilize surgical tools require a temperature of 120°C to kill bacteria. If water is used for this purpose, at what pressure must the autoclave operate? The normal boiling point of water is 373.15 K, and $\Delta H^\circ_{vaporization} = 40.656\times10^3\text{ J mol}^{-1}$ at the normal boiling point.

$$\ln\frac{P_f}{P_i} = -\frac{\Delta H_m^{vaporization}}{R}\left(\frac{1}{T_f} - \frac{1}{T_i}\right)$$

$$\ln\frac{P_f}{P_i} = -\frac{40.656\times10^3\text{J mol}^{-1}}{8.314\text{ J mol}^{-1}\text{K}^{-1}} \times \left(\frac{1}{393.15\text{ K}} - \frac{1}{373.15\text{ K}}\right) = 0.6667$$

$$\frac{P_f}{P_i} = 1.95; \quad P_f = 1.95\text{ atm}$$

P8.33) Calculate the factor by which vapor pressure of a droplet of methanol of radius 1.00×10^{-4} m at 45.0° C in equilibrium with its vapor is increased with respect to a very large droplet. Use the tabulated value of the density and the surface tension at 298 K from Appendix A for this problem. (*Hint:* You need to calculate the vapor pressure of methanol at this temperature.)

$$\ln\frac{P(T)}{\text{Pa}} = A(1) - \frac{A(2)}{\frac{T}{\text{K}} + A(3)} = 23.593 - \frac{3.6971\times10^{3}}{325 - 31.317} = 11.0043$$

$$P = 6.01\times10^{4}\,\text{Pa}$$

$$\Delta P = \frac{2\gamma}{r} = \frac{2\times22.07\times10^{-3}\,\text{N}\,\text{m}^{-1}}{10^{-8}\text{m}} = 4.41\times10^{6}\ \text{Pa}$$

$$P_{inside} = P_{vapor} + \Delta P = 6.01\times10^{4}\text{Pa} + 4.41\times10^{6}\ \text{Pa} = 4.47\times10^{6}\text{Pa}$$

For a very large droplet, $\Delta P \rightarrow 0$, and the vapor pressure is $6.01\times10^{4}\,\text{Pa}$. For the small droplet, the vapor pressure is increased by the factor

$$\ln\left(\frac{P}{P_0}\right) = \frac{\frac{\rho}{M}(\mathbf{P} - P_0)}{RT} = \frac{\frac{32.04\times10^{-3}\text{kg}\,\text{mol}^{-1}}{791.4\ \text{kg}\ \text{m}^{-3}}\times\left(4.47\times10^{6} - 6.01\times10^{4}\right)\text{Pa}}{8.314\ \text{J}\,\text{mol}^{-1}\text{K}^{-1}\times325\ \text{K}} = 6.607\times10^{-2}$$

$$P = 1.068P_0$$

$$h = \frac{(101325 - 3242)\,\text{Pa}}{998\ \text{kg}\ \text{m}^{-3}\times9.81\ \text{m}\ \text{s}^{-2}} = 10.0\ \text{m}$$

Chapter 9: Ideal and Real Solutions

P9.1) At 303 K, the vapor pressure of benzene is 118 Torr and that of cyclohexane is 122 Torr. Calculate the vapor pressure of a solution for which $x_{benzene} = 0.25$ assuming ideal behavior.

$$P_{total} = x_{benzene} P^*_{benzene} + x_{cyclohexene} P^*_{cyclohexene}$$

$$= 0.25 \times 118 \text{ Torr} + (1 - 0.25) \times 122 \text{ Torr} = 121 \text{ Torr}$$

P9.3) An ideal solution is formed by mixing liquids A and B at 298 K. The vapor pressure of pure A is 180 Torr and that of pure B is 82.1 Torr. If the mole fraction of A in the vapor is 0.450, what is the mole fraction of A in the solution?

$$x_A = \frac{y_A P_B^*}{P_A^* + (P_B^* - P_A^*) y_A} = \frac{0.450 \times 82.1 \text{ Torr}}{180 \text{ Torr} + (82.1 \text{ Torr} - 180 \text{ Torr}) \times 0.450} = 0.272$$

P9.5) A and B form an ideal solution at 298 K, with $x_A = 0.600$, $P_A^* = 105$ Torr and $P_B^* =$ 63.5 Torr.
a) Calculate the partial pressures of A and B in the gas phase.
b) A portion of the gas phase is removed and condensed in a separate container. Calculate the partial pressures of A and B in equilibrium with this liquid sample at 298 K.

a) Calculate the partial pressures of A and B in the gas phase.

$$P_A = x_A P_A^* = 0.600 \times 105 \text{ Torr} = 63.0 \text{ Torr}$$

$$P_B = (1 - x_A) P_B^* = 0.400 \times 63.5 \text{ Torr} = 25.4 \text{ Torr}$$

b) A portion of the gas phase is removed and condensed in a separate container. Calculate the partial pressures of A and B in equilibrium with this liquid sample at 298 K. The composition of the initial gas is given by

$$y_A = \frac{P_A}{P_A + P_B} = \frac{63.0 \text{ Torr}}{88.4 \text{ Torr}} = 0.71; \quad y_B = 0.287$$

For the portion removed, the new x_A and x_B values are the previous y_A and y_B values.

$$P_A = x_A P_A^* = 0.713 \times 105 \text{ Torr} = 74.9 \text{ Torr}$$

$$P_B = (1 - x_A) P_B^* = 0.287 \times 63.5 \text{ Torr} = 18.2 \text{ Torr}$$

P9.6) The vapor pressures of 1-bromobutane and 1-chlorobutane can be expressed in the form $\ln \frac{P_{bromo}}{\text{Pa}} = 17.076 - \frac{1584.8}{\frac{T}{\text{K}} - 111.88}$ and $\ln \frac{P_{chloro}}{\text{Pa}} = 20.612 - \frac{2688.1}{\frac{T}{\text{K}} - 55.725}$. Assuming ideal solution behavior, calculate x_{bromo} and y_{bromo} at 300.0 K and a total pressure of 8741 Pa.

At 300.0K, $P^\circ_{bromo} = 5719$ Pa and $P^\circ_{chloro} = 14877$ Pa.

$$P_{total} = x_{bromo}P^{\circ}_{bromo} + (1 - x_{bromo})P^{\circ}_{chloro}$$

$$x_{bromo} = \frac{P_{total} - P^{\circ}_{chloro}}{P^{\circ}_{bromo} - P^{\circ}_{chloro}} = \frac{8741 \text{ Pa} - 14877 \text{ Pa}}{5719 \text{ Pa} - 14877 \text{ Pa}} = 0.67$$

$$y_{bromo} = \frac{x_{bromo}P^{\circ}_{bromo}}{P_{total}} = \frac{0.67 \times 5719 \text{ Pa}}{8741 \text{ Pa}} = 0.44$$

P9.8) An ideal solution at 298 K is made up of the volatile liquids A and B, for which P_A^*=125 Torr and P_B^*= 46.3 Torr. As the pressure is reduced from 450 Torr, the first vapor is observed at a pressure of 70.0 Torr. Calculate x_A.
The first vapor is observed at a pressure of

$$P_{total} = x_A P_A^* + (1 - x_A)P_B^*$$

$$x_A = \frac{P_{total} - P_B^*}{P_A^* - P_B^*} = \frac{70.0 \text{ Torr} - 46.3 \text{ Torr}}{125 \text{ Torr} - 46.3 \text{ Torr}} = 0.301$$

P9.10) At –31.2°C, pure propane and *n*-butane have a vapor pressure of 1200 and 200 Torr, respectively.
a) Calculate the mole fraction of propane in the liquid mixture that boils at –31.2°C at a pressure of 760 Torr.
b) Calculate the mole fraction of propane in the vapor that is in equilibrium with the liquid of part (a).

a) $P_{total} = x_p P_p^* + (1 = x_p)P_B^*$

$$760 \text{ Torr} = 1200x_p + 200(1 - x_p)$$

$$x_p = 0.560$$

b) $$y_B = \frac{x_p P_p^*}{P_B^* + (P_p^* - P_B^*)x_p}$$

$$= \frac{0.56 \times 1200 \text{ Torr}}{200 \text{ Torr} + 0.56 \times (1200 \text{ Torr} - 200 \text{ Torr})}$$

$$= 0.884$$

P9.11) In an ideal solution of A and B, 3.50 mol are in the liquid phase and 4.75 mol are in the gaseous phase. The overall composition of the system is Z_A = 0.300 and x_A = 0.250. Calculate y_A.

$$n_{liq}^{tot}(Z_B - x_B) = n_{vapor}^{tot}(y_B - Z_B)$$

$$y_B = \frac{n_{liq}^{tot}(Z_B - x_B) + n_{vapor}^{tot} Z_B}{n_{vapor}^{tot}} = \frac{3.50 \text{ mol} \times (0.300 - 0.250) + 4.75 \text{ mol} \times 0.300}{4.75 \text{ mol}} = 0.337$$

P9.13) At 39.9°C, a solution of ethanol ($x_1 = 0.9006$, $P_1^* = 130.4$ Torr) and isooctane ($P_2^* = 43.9$ Torr) forms a vapor phase with $y_1 = 0.6667$ at a total pressure of 185.9 Torr.
a) Calculate the activity and activity coefficient of each component.
b) Calculate the total pressure that the solution would have if it were ideal.

a) The activity and activity coefficient for ethanol are given by

$$a_1 = \frac{y_1 P_{total}}{P_1^*} = \frac{0.6667 \times 185.9 \text{ Torr}}{130.4 \text{ Torr}} = 0.9504$$

$$\gamma_1 = \frac{a_1}{x_1} = \frac{0.9504}{0.9006} = 1.055$$

Similarly, the activity and activity coefficient for isooctane are given by

$$a_2 = \frac{(1 - y_1) P_{total}}{P_2^*} = \frac{0.3333 \times 185.9 \text{ Torr}}{43.9 \text{ Torr}} = 1.411$$

$$\gamma_2 = \frac{a_2}{x_2} = \frac{1.411}{1 - 0.9006} = 14.20$$

b) If the solution were ideal, Raoult's law would apply.

$$P_{Total} = x_1 P_1^* + x_2 P_2^*$$
$$= 0.9006 \times 130.4 \text{ Torr} + (1 - 0.9006) \times 43.9 \text{ Torr}$$
$$= 121.8 \text{ Torr}$$

P9.19) A solution is prepared by dissolving 32.5 g of a nonvolatile solute in 200 g of water. The vapor pressure above the solution is 21.85 Torr and the vapor pressure of pure water is 23.76 Torr at this temperature. What is the molecular weight of the solute?

$$x_{H_2O} = \frac{P_{H_2O}}{P_{H_2O}^*} = \frac{21.85 \text{ Torr}}{23.76 \text{ Torr}} = 0.9196$$

$$x_{solute} = 0.0804 = \frac{n_{solute}}{n_{solute} + n_{H_2O}};$$

$$n_{solute} = \frac{0.0804\ n_{H_2O}}{0.9617} = \frac{0.0804 \times \dfrac{200 \text{ g}}{18.02 \text{ g mol}^{-1}}}{0.9196} = 0.970 \text{ mol}$$

$$M = \frac{32.5 \text{ g}}{0.970 \text{ mol}} = 33.5 \text{ g mol}^{-1}$$

P9.21) The dissolution of 5.25 g of a substance in 565 g of benzene at 298 K raises the boiling point by 0.625°C. Note that $K_f = 5.12$ K kg mol^{-1}, $K_b = 2.53$ K kg mol^{-1}, and the density of benzene is 876.6 kg m^{-3}. Calculate the freezing point depression, the ratio of the vapor pressure above the solution to that of the pure solvent, the osmotic pressure, and the molecular weight of the solute. $P_{benzene}^* = 103$ Torr at 298 K.

$$\Delta T_b = K_b m_{solute};\quad m_{solute} = \frac{\Delta T_b}{K_b} = \frac{0.625\ \text{K}}{2.53\ \text{K kg mol}^{-1}} = 0.247\ \text{mol kg}^{-1}$$

$$M = \frac{5.25\ \text{g}}{0.247\ \text{mol kg}^{-1} \times 0.565\ \text{kg}} = 37.6\ \text{g mol}^{-1}$$

$$\Delta T_f = -K_f m_{solute} = -5.12\ \text{K kg mol}^{-1} \times 0.247\ \text{mol kg}^{-1} = -1.26\ \text{K}$$

$$\frac{P_{benzene}}{P^*_{benzene}} = x_{benzene} = \frac{n_{benzene}}{n_{benzene} + n_{solute}}$$

$$= \frac{\dfrac{565\ \text{g}}{78.11\ \text{g mol}^{-1}}}{\dfrac{565\ \text{g}}{78.11\ \text{g mol}^{-1}} + 0.247\ \text{mol kg}^{-1} \times 0.565\ \text{kg}} = 0.981$$

$$\pi = \frac{n_{solute}RT}{V} = \frac{\dfrac{5.25\times10^{-3}\text{kg}}{37.6\times10^{-3}\text{kg mol}^{-1}} \times 8.314\ \text{J mol}^{-1}\text{K}^{-1} \times 298\ \text{K}}{\dfrac{565\times10^{-3}\text{kg}}{876.6\text{kg m}^{-3}}} = 5.37\times10^5\text{Pa}$$

P9.26) Calculate the solubility of H_2S in 1 L of water if its pressure above the solution is 3.25 bar. The density of water at this temperature is 997 kg m^{-3}.

$$x_{H_2S} = \frac{n_{H_2S}}{n_{H_2S} + n_{H_2O}} \approx \frac{n_{H_2S}}{n_{H_2O}} = \frac{P_{H_2S}}{k^H_{H_2S}} = \frac{3.25\ \text{bar}}{568\ \text{bar}} = 5.72\times10^{-3}$$

$$n_{H_2O} = \frac{\rho_{H_2O}V}{M_{H_2O}} = \frac{10^{-3}\text{m}^3 \times 997\ \text{kg m}^{-3}}{18.02\times10^{-3}\text{kg mol}^{-1}} = 55.4$$

$$n_{H_2S} = x_{H_2S}n_{H_2O} = 5.72\times10^{-3} \times 55.4 = 0.327\ \text{mol}$$

P9.28) At a given temperature, a nonideal solution of the volatile components A and B has a vapor pressure of 832 Torr. For this solution, $y_A = 0.404$. In addition, $x_A = 0.285$, $P^*_A = 591$ Torr, and $P^*_B = 503$ Torr. Calculate the activity and activity coefficient of A and B.

$$P_A = y_A P_{total} = 0.404 \times 832\ \text{Torr} = 336\ \text{Torr}$$

$$P_B = 832\ \text{Torr} - 336\ \text{Torr} = 496\ \text{Torr}$$

$$a_A = \frac{P_A}{P^*_A} = \frac{336\ \text{Torr}}{591\ \text{Torr}} = 0.569$$

$$\gamma_A = \frac{a_A}{x_A} = \frac{0.569}{0.285} = 2.00$$

$$a_B = \frac{P_B}{P^*_B} = \frac{496\ \text{Torr}}{503\ \text{Torr}} = 0.986$$

$$\gamma_B = \frac{a_B}{x_B} = \frac{0.986}{0.715} = 1.38$$

Chapter 10: Electrolyte Solutions

P10.1) Calculate $\Delta H^\circ_{reaction}$ and $\Delta G^\circ_{reaction}$ for the reaction $AgNO_3(aq) + KCl(aq) \rightarrow AgCl(s) + KNO_3(aq)$.

$$\Delta G^\circ_{reaction} = \Delta G^\circ_f(\text{AgCl}, s) + \Delta G^\circ_f(\text{K}^+, aq) + \Delta G^\circ_f(\text{NO}_3^-, aq) - \Delta G^\circ_f(\text{Ag}^+, aq) - \Delta G^\circ_f(\text{NO}_3^-, aq) - \Delta G^\circ_f(\text{K}^+, aq) - \Delta G^\circ_f(\text{Cl}^-, aq)$$

$$\Delta G^\circ_{reaction} = \Delta G^\circ_f(\text{AgCl}, s) - \Delta G^\circ_f(\text{Ag}^+, aq) - \Delta G^\circ_f(\text{Cl}^-, aq) = -109.8 \text{ kJ mol}^{-1} - 77.1 \text{ kJ mol}^{-1} + 131.2 \text{ kJ mol}^{-1} = -55.7 \text{ kJ mol}^{-1}$$

$$\Delta H^\circ_{reaction} = \Delta H^\circ_f(\text{AgCl}, s) - \Delta H^\circ_f(\text{Ag}^+, aq) - \Delta H^\circ_f(\text{Cl}^-, aq)$$
$$\Delta H^\circ_{reaction} = -127.0 \text{ kJ mol}^{-1} - 105.6 \text{ kJ mol}^{-1} + 167.2 \text{ kJ mol}^{-1} = -65.4 \text{ kJ mol}^{-1}$$

P10.3) Calculate $\Delta S^\circ_{reaction}$ for the reaction $AgNO_3(aq) + KCl(aq) \rightarrow AgCl(s) + KNO_3(aq)$.

$$\Delta S^\circ_{reaction} = S^\circ(\text{AgCl}, s) - S^\circ(\text{Ag}^+, aq) - S^\circ(\text{Cl}^-, aq)$$
$$\Delta S^\circ_{reaction} = 96.3 \text{ J K}^{-1}\text{ mol}^{-1} - 72.7 \text{ J K}^{-1}\text{ mol}^{-1} - 56.5 \text{ J K}^{-1}\text{ mol}^{-1} = -32.9 \text{ J K}^{-1}\text{ mol}^{-1}$$

P10.7) Express $\mu_\pm$ in terms of μ_+ and μ_- for a) NaCl, b) $MgBr_2$, c) Li_3PO_4, and d) $Ca(NO_3)_2$. Assume complete dissociation.

a) NaCl $\mu_\pm = \dfrac{\mu_{solute}}{v} = \dfrac{v_+\mu_+ + v_-\mu_-}{v} = \dfrac{\mu_+ + \mu_-}{2}$

b) $MgBr_2$ $\mu_\pm = \dfrac{\mu_{solute}}{v} = \dfrac{v_+\mu_+ + v_-\mu_-}{v} = \dfrac{\mu_+ + 2\mu_-}{3}$

c) Li_3PO_4 $\mu_\pm = \dfrac{\mu_{solute}}{v} = \dfrac{v_+\mu_+ + v_-\mu_-}{v} = \dfrac{3\mu_+ + \mu_-}{4}$

d) $Ca(NO_3)_2$ $\mu_\pm = \dfrac{\mu_{solute}}{v} = \dfrac{v_+\mu_+ + v_-\mu_-}{v} = \dfrac{\mu_+ + 2\mu_-}{3}$

P10.10) Calculate the ionic strength in a solution that is 0.0050*m* in K_2SO_4, 0.0010*m* in Na_3PO_4, and 0.0025*m* in $MgCl_2$.

$$I_{K_2SO_4} = \frac{m}{2}\left(v_+z_+^2 + v_-z_-^2\right)$$
$$= \frac{0.0050}{2}(2+4) = 0.0150 \text{ mol kg}^{-1}$$

$$I_{Na_3PO_4} = \frac{m}{2}\left(v_+ z_+^2 + v_- z_-^2\right)$$
$$= \frac{0.0010}{2}(3+9) = 0.0060 \text{ mol kg}^{-1}$$

$$I_{MgCl_2} = \frac{m}{2}\left(v_+ z_+^2 + v_- z_-^2\right)$$
$$= \frac{0.0025}{2}(4+2) = 0.0075 \text{ mol kg}^{-1}$$

total ionic strength

$I = (0.0150 + 0.0060 + 0.0075)$ mol kg^{-1}
$= 0.0285$ mol kg^{-1}

P10.14) Calculate the Debye–Hückel screening length $\frac{1}{\kappa}$ at 298 K in a 0.00100*m* solution of NaCl.

$$\kappa = 9.211\times10^{8}\sqrt{\frac{I/\text{mol kg}^{-1}}{\varepsilon_r}}\ \text{m}^{-1} = 9.211\times10^{8}\sqrt{\frac{0.00100}{78.54}}\ \text{m}^{-1} = 3.29\times10^{6}\,\text{m}^{-1}$$

$$\frac{1}{\kappa} = 3.04\times10^{-7}\,\text{m} = 304 \text{ nm}$$

P10.17) Calculate I, $\gamma_\pm$, and $a_\pm$ for a 0.0250*m* solution of $AlCl_3$ at 298 K. Assume complete dissociation.

$$\text{AlCl}_3 \Rightarrow v_+ = 1,\ v_- = 3,\ z_+ = 3,\ z_- = 1$$
$$I = \frac{0.0250}{2}(9+3) = 0.1500 \text{ mol kg}^{-1}$$
$$\ln\gamma_\pm = -1.173\times3\times\sqrt{0.1500} = -1.3629$$
$$\gamma_\pm = 0.2559$$
$$a_\pm = \left(\frac{m_\pm}{m^\circ}\right)\gamma_\pm$$
$$m_\pm^4 = (0.025)(0.025\times3)^3 = 1.0546875\times10^{-5}$$
$$m_\pm = 0.05699 \text{ mol kg}^{-1}$$
$$a_\pm = 0.05699\times0.2559 = 0.0146$$

P10.20) Calculate the solubility of $BaSO_4$ ($K_{sp} = 1.08 \times 10^{-10}$) (a) in pure H_2O and (b) in an aqueous solution with $I = 0.0010$ mol kg^{-1}.

a) $BaSo_4(s) \rightarrow Ba^{2+}(aq) + SO_4^{2-}(aq)$

$$\nu_+ = 1, \quad \nu_- = 1$$

$$z_+ = 2, \quad z_- = 2$$

$$K_{sp} = \left(\frac{c_{B_a^{2+}}}{c^\circ}\right)\left(\frac{c_{SO_4^{2-}}}{c^\circ}\right)\gamma_\pm^2 = 1.08 \times 10^{-10}$$

$$c_{Ba^{2+}} = c_{SO_4^{2-}}$$

$$K_{sp} = \left(\frac{C_{Ba^{2+}}}{C^\circ}\right)^2 \gamma_\pm^2 = 1.08 \times 10^{-10}$$

when $\gamma_\pm = 1$ $\quad c_{Ba^{2+}} = 1.039 \times 10^{-5}$ mol L^{-1}

$$I = \frac{m}{2}\sum\left(\nu_+ z_+^2 + \nu_- z_-^2\right)$$

$$= \frac{1.039 \times 10^{-5}}{2} \times (4+4) = 4.157 \times 10^{-5} \text{ mol kg}^{-1}$$

$$\ln \gamma_\pm = -1.173 \times 4 \times \sqrt{4.157 \times 10^{-5}} = -0.03025$$

$$\gamma_\pm = 0.97020$$

when $\gamma_\pm = 0.970230$ $\quad c_{Ba^{2+}} = 1.0711 \times 10^{-5}$ mol L^{-1}

$$I = \frac{1.0711 \times 10^{-5}}{2} \times (8) = 4.2846 \times 10^{-5} \text{ mol kg}^{-1}$$

$$\ln \gamma_\pm = -1.173 \times 4 \times \sqrt{4.2846 \times 10^{-5}} = -0.03071$$

$$\gamma_\pm = 0.9698$$

when $\gamma_\pm = 0.9698$ $\quad c_{Ba^{2+}} = 1.0716 \times 10^{-5}$ mole L^{-1}

$$I = \frac{1.0716 \times 10^{-5}}{2} \times (8) = 4.2866 \times 10^{-5} \text{ mol kg}^{-1}$$

$$\ln \gamma_\pm = -1.173 \times 4 \times \sqrt{4.2866 \times 10^{-5}} = -0.03072$$

$$\gamma_\pm = 0.9697$$

$$\text{solubility} = 1.07 \times 10^{-5} \text{ mol L}^{-1}$$

b) $I = 0.0010 \text{ mol kg}^{-1}$

$$\ln \gamma_\pm = 1.173 \times 4 \times \sqrt{0.0010} = -0.148374$$

$$\gamma_\pm = 0.8621$$

$$K_{sp} = \left(\frac{c_{Ba^{2+}}}{c^\circ}\right) \times (0.8621)^2 = 1.08 \times 10^{-10}$$

$$c_{Ba^{2+}} = 1.21 \times 10^{-5} \text{ mol kg}^{-1}$$

The relative error is 20%. There is no need for a further iteration because the ionic strength in the solution is not influenced by the dissociation of the $BaSO_4$.

P10.21) Dichloroacetic acid has a dissociation constant of $K_a = 3.32 \times 10^{-2}$. Calculate the degree of dissociation for a $0.125m$ solution of this acid (a) using the Debye–Hückel limiting law and (b) assuming that the mean ionic activity coefficient is one.

a) We first consider the case when γ is given by the Debye–Hückel limiting law. The ionic strength is given by

$$I = \frac{m}{2}(2) = m = 0.04992 \text{ mol kg}^{-1}$$

$$\ln \gamma_\pm = -1.173 \times 1 \times \sqrt{0.04992} = -0.2621$$

$$\gamma_\pm = 0.7694$$

We recalculate the ionic strength and iterate several times.

$$I = \frac{m}{2}(2) = m = 0.04992 \text{ mol kg}^{-1}$$

$$\ln \gamma_\pm = -1.173 \times 1 \times \sqrt{0.04992} = -0.2621$$

$$\gamma_\pm = 0.7694$$

when $\gamma_\pm = 0.7694$, $\dfrac{m^2 \gamma_\pm^2}{0.125 - m} = 3.32 \times 10^{-2}$

$$m^2 = 7.0097 \times 10^{-3} - 0.05608 \text{ m}$$

$$m^2 + 0.05608 \text{ m} - 7.0097 \times 10^{-3} = 0$$

$$m = \frac{-0.05608 \pm \sqrt{(0.05608)^2 - 4(1)(-7.0097 \times 10^{-3})}}{2} = 0.06025 \text{ mol kg}^{-1}$$

$$I = 0.06025 \text{ mol kg}^{-1}$$

$$\ln \gamma_\pm = -1.173 \times 1 \times \sqrt{0.06025} = -0.2879$$

$$\gamma_\pm = 0.7498$$

when $\gamma_{\pm} = 0.7498$

$$\frac{m^2}{0.125 - m} = 0.05905$$

$$m^2 + 0.05905\ \text{m} - 7.3815\times10^{-3} = 0$$

$$m = \frac{-0.05905 \pm \sqrt{(0.05905)^2 - 4(1)(-7.3815\times10^{-3})}}{2} = 0.06132\ \text{mol kg}^{-1}$$

$$I = 0.06132\ \text{mol kg}^{-1}$$

$$\ln\gamma_{\pm} = 0.7479$$

$$\gamma_{\pm} = 0.7479$$

when $\gamma_{\pm} = 0.7479$

$$\frac{m^2}{0.125 - m} = 0.05935$$

$$m^2 + 0.05935\ \text{m} - 7.4191\times10^{-3} = 0$$

$$m = \frac{-0.05935 \pm \sqrt{(0.05935)^2 - 4(1)(-7.4191\times10^{-3})}}{2} = 0.06143\ \text{mol kg}^{-1}$$

$$I = 0.06143\ \text{mol kg}^{-1}$$

$$\ln\gamma_{\pm} = -0.2907$$

$$\gamma_{\pm} = 0.7477$$

when $\gamma_{\pm} = 0.7477$

$$\frac{m^2}{0.125 - m} = 0.05938$$

$$m^2 + 0.05938\ \text{m} - 7.4228\times10^{-3} = 0$$

$$m = \frac{-0.05938 \pm \sqrt{(0.05938)^2 - 4(1)(-7.4228\times10^{-3})}}{2} = 0.06144\ \text{mol kg}^{-1}$$

This is a sufficiently good result, and we calculate the degree of dissociation to be $\frac{0.06144}{0.125}\times100\% = 49\%$.

b) $\gamma_\pm = 1$

$$m^2 = 4.15\times10^{-3} - 3.32\times10^{-2}\ \text{m}$$
$$m^2 + 3.32\times10^{-2}\ \text{m} - 4.15\times10^{-3} = 0$$
$$m = \frac{-3.32\times10^{-2} \pm \sqrt{\left(3.32\times10^{-2}\right)^2 - 4(1)\left(-4.15\times10^{-3}\right)}}{2} = 0.04992\ \text{mol kg}^{-1}$$
$$\frac{0.04992}{0.125}\times100\% = 40\%$$

This result is 20% smaller than that calculated using the Debye–Hückel limiting law.

P10.23) The equilibrium constant for the hydrolysis of dimethylamine

$$(CH_3)_2NH(aq) + H_2O(aq) \rightarrow CH_3NH_3^+(aq) + OH^-(aq)$$

is 5.12×10^{-4}. Calculate the extent of hydrolysis for (a) a 0.125*m* solution of $(CH_3)_2NH$ in water and (b) a solution that is also 0.045*m* in $NaNO_3$.

a) $\left(CH_3\right)_2 NH + H_2O \rightarrow CH_3NH_3^+ + OH^-$

$$K = \frac{m^2\gamma_\pm^2}{0.125 - m} = 5.12\times10^{-4}$$

if $\gamma_\pm = 1$,

$m = 7.75\times10^{-3}\ \text{mol kg}^{-1}$

when $\gamma_\pm = 1$

$$I = \frac{m}{2}(2) = m = 7.75\times10^{-3}\ \text{mol kg}^{-1}$$
$$\ln\gamma_\pm = -1.173\times1\times\sqrt{7.75\times10^{-3}} = -0.1033$$
$$\gamma_\pm = 0.9019$$

when $\gamma_\pm = 0.9019$

$$m = 8.561\times10^{-3}\ \text{mol kg}^{-1}$$
$$I = m = 8.561\times10^{-3}\ \text{mol kg}^{-1}$$
$$\ln\gamma_\pm = -0.1085$$
$$\gamma_\pm = 0.8971$$

when $\gamma_\pm = 0.8971$

$$m = 8.605\times10^{-3}\ \text{mol kg}^{-1}$$

$$I = m = 8.605\times10^{-3}\ \text{mol kg}^{-1}$$

$$\ln\gamma_\pm = -0.1088$$

$$\gamma_\pm = 0.8969$$

when $\gamma_\pm = 0.8969$

$$m = 8.61\times10^{-3}\ \text{mol kg}^{-1}$$

$$\frac{8.61\times10^{-3}}{0.125}\times100\% = 6.89\%$$

b) $NaNO_3$

$$\begin{array}{llll} Na^+ & NO_3^- & \nu_+ = 1 & \nu_- = 1 \\ 0.045 & 0.045 & z_+ = 1 & z_- = 1 \end{array}$$

$$I = \frac{0.045}{2}(1+1) = 0.045\ \text{mol kg}^{-1}$$

Add to this the ionic strength from part (a).

$$I_{total} = 0.045 + 8.605\times10^{-3} = 0.0536\ \text{mol kg}^{-1}$$

$$\ln\gamma_\pm = -1.173\times1\times\sqrt{0.0536} = -0.2716$$

$$\gamma_\pm = 0.7622$$

$$K = \frac{(m)^2(\gamma_\pm)^2}{0.125} = \frac{(m)^2(0.7622)^2}{0.125} = 5.12\times10^{-4}$$

$$m = 0.01006\ \text{mol kg}^{-1}$$

Carrying out another iteration

$$I_{total} = 0.045 + .01006 = 0.0551\ \text{mol kg}^{-1}$$

$$\ln\gamma_\pm = -1.173\times1\times\sqrt{0.0551} = -0.2753$$

$$\gamma_\pm = 0.7593$$

$$K = \frac{(m)^2(\gamma_\pm)^2}{0.125} = \frac{(m)^2(0.7593)^2}{0.125} = 5.12\times10^{-4}$$

$$m = 0.01010\ \text{mol kg}^{-1}$$

The degree of dissociation is

$$\frac{0.01010}{0.125}\times100\% = 8.08\%$$

P10.25) Calculate the mean ionic molality, $m_{\pm}$, in $0.0500m$ solutions of (a) $Ca(NO_3)_2$, (b) NaOH, (c) $MgSO_4$, and (d) $AlCl_3$.

$$m_{\pm}^{\nu} = m_{+}^{\nu_+} m_{-}^{\nu_-}$$

a) $Ca(NO_3)_2 \quad \nu_+ = 1, \ \nu_- = 2$

$$m_{\pm}^{3} = (0.0500)(0.0500\times 2)^2 = 5.00\times 10^{-4}$$

$$m_{\pm} = 0.0794 \text{ mol kg}^{-1}$$

b) NaOH $\quad \nu_+ = 1, \ \nu_- = 1, \ \nu = 2$

$$m_{\pm}^{2} = (0.0500)(0.0500)$$

$$m_{\pm} = 0.0500 \text{ mol kg}^{-1}$$

c) $MgSO_4 \quad \nu_+ = 1, \ \nu_- = 1, \ \nu = 2$

$$m_{\pm}^{2} = (0.0500)(0.0500)$$

$$m_{\pm} = 0.0500 \text{ mol kg}^{-1}$$

d) $AlCl_3 \quad \nu_+ = 1, \ \nu_- = 3, \ \nu = 4$

$$m_{\pm}^{4} = (0.0500)(0.0500\times 3)^3 = 1.6875\times 10^{-4}$$

$$m_{\pm} = 0.1140 \text{ mol kg}^{-1}$$

P10.27) At 25°C, the equilibrium constant for the dissociation of acetic acid, K_a, is 1.75×10^{-5}. Using the Debye–Hückel limiting law, calculate the degree of dissociation in $0.100m$ and $1.00m$ solutions. Compare these values with what you would obtain if the ionic interactions had been ignored.

$$CH_3COOH \rightarrow CH_3COO^- + H^+$$

For 0.100 m

$$\frac{m^2 \gamma_{\pm}^2}{0.100 - m} = 1.75\times 10^{-5}$$

when $\gamma_{\pm} = 1$

$m = 1.314\times10^{-3}\ \text{mol kg}^{-1}$

$I = \frac{m}{2}(2) = m = 1.314\times10^{-3}$

$\ln\gamma_{\pm} - 1.173\times1\times\sqrt{1.314\times10^{-3}} = -0.04252$

$\gamma_{\pm} = 0.9584$

when $\gamma_{\pm} = 0.9584$

$m = 1.38056\times10^{-3}\ \text{mol kg}^{-1}$

$I = m = 1.371\times10^{-3}\ \text{mol kg}^{-1}$

$\ln\gamma_{\pm} = -0.04343$

$\gamma_{\pm} = 0.9575$

when $\gamma_{\pm} = 0.9575$

$m = 1.372\times10^{-3}\ \text{mol kg}^{-1}$

$I = m = 1.372\times10^{-3}\ \text{mol kg}^{-1}$

$\ln\gamma_{\pm} = -0.04345$

$\gamma_{\pm} = 0.9575$

This result has converged sufficiently to calculate the degree of dissociation.

$m = 1.372\times10^{-3}$

$$\frac{1.372\times10^{-3}}{0.100}\times100\% = 1.37\%$$

for 1.00 m

$$\frac{m^2\gamma_{\pm}^2}{1.00} = 1.75\times10^{-5}$$

$m^2\gamma_{\pm}^2 = 1.75\times10^{-5}$

$m\gamma_{\pm} = 4.1833\times10^{-3}$

when $\gamma_{\pm} = 1$

$m = 4.1833\times10^{-3}\ \text{mol kg}^{-1}$

$I = m = 4.1833\times10^{-3}\ \text{mol kg}^{-1}$

$\ln\gamma_{\pm} = -0.07587$

$\gamma_{\pm} = 0.9269$

when $\gamma_{\pm} = 0.9269$

$m = 4.5130 \times 10^{-3}\ \mathrm{mol\,kg^{-1}}$

$I = m = 4.5130 \times 10^{-3}\ \mathrm{mol\,kg^{-1}}$

$\ln \gamma_{\pm} = -0.07880$

$\gamma_{\pm} = 0.9242$

when $\gamma_{\pm} = 0.9242$

$m = 4.5263 \times 10^{-3}\ \mathrm{mol\,kg^{-1}}$

$I = m = 4.5263 \times 10^{-3}\ \mathrm{mol\,kg^{-1}}$

$\ln \gamma_{\pm} = -0.07892$

$\gamma_{\pm} = 0.9241$

when $\gamma_{\pm} = 0.9241$

$m = 4.5268 \times 10^{-3}\ \mathrm{mol\,kg^{-1}}$

$I = m = 4.5268 \times 10^{-3}\ \mathrm{mol\,kg^{-1}}$

$\ln \gamma_{\pm} = -0.07892$

$\gamma_{\pm} = 0.9241$

$$\frac{4.5268 \times 10^{-3}}{1.0} \times 100\% = 0.453\%$$

If ionic interactions are ignored

for 0.100 m

$$\frac{m^2}{0.100 - m} = 1.75 \times 10^{-5}$$

$m = \sqrt{1.75 \times 10^{-6}} = 1.314 \times 10^{-3}\ \mathrm{mol\,kg^{-1}}$

$$\frac{1.314 \times 10^{-3}}{0.100} \times 100\% = 1.31\%$$

for 1.00 m

$$\frac{m^2}{0.10} = 1.75 \times 10^{-5}$$

$m = \sqrt{1.75 \times 10^{-5}} = 4.1833 \times 10^{-3}\ \mathrm{mol\,kg^{-1}}$

$$\frac{4.1833 \times 10^{-3}}{0.10} \times 100\% = 0.418\%$$

Chapter 11: Electrochemical Cells, Batteries, and Fuel Cells

P11.8) Consider the cell $Hg(l)|Hg_2SO_4(s)|FeSO_4(aq, a = 0.0100)|Fe(s)$.

a) Write the cell reaction.

b) Calculate the cell potential, the equilibrium constant for the cell reaction, and ΔG° at 25°C.

a) Oxidation: $2Hg + SO_4^{2-} \rightarrow Hg_2SO_4 + 2e^- \ (1) \qquad E^\circ = -0.6125 \text{ V}$

Reduction: $Fe^{2+} + 2e^- \rightarrow Fe(2) \qquad E^\circ = -0.447 \text{ V}$

Cell reaction:

$$2Hg(l) + Fe^{2+}(aq) + SO_4^{2-}(aq) \rightarrow Hg_2SO_4(s) + Fe(s)$$

$$E^\circ_{cell} = -0.6125 + (-0.447) = -1.0595 \text{ V}$$

$$E_{cell} = E^\circ_{cell} - \frac{RT}{nF}\ln\left(\frac{1}{a_{Fe^{2+}}a_{SO_4^{2-}}}\right)$$

$$= -1.0595 - \frac{(8.3145 \text{ J mol}^{-1}\text{K}^{-1})\times(298.15 \text{ K})}{2\times(96,485 \text{ C mol}^{-1})}\ln\left(1/(0.01)^2\right)$$

$$= 1.178 \text{ V}$$

$$\Delta G^\circ = -nFE^\circ = (-2)(96,485 \text{ C mol}^{-1})(-1.0595 \text{ V})$$

$$= 204.5 \text{ kJ mol}^{-1}$$

$$K = e^{-\Delta G^\circ/RT} = 1.49\times10^{-36}$$

P11.14) Determine E° for the reaction $Cr^{2+} + 2e^- \rightarrow Cr$ from the one-electron reduction potential for Cr^{3+} and the three-electron reduction potential for Cr^{3+} given in Table 11.1 (see Appendix B).

$Cr^{3+} + 3e^- \rightarrow Cr \qquad \Delta G^\circ = -nFE^\circ = -3\times96485 \text{ C mol}^{-1}\times(-0.744 \text{ V}) = 215.4 \text{ kJ mol}^{-1}$

$Cr^{2+} \rightarrow Cr^{3+} + e^- \qquad \Delta G^\circ = -nFE^\circ = -1\times96485 \text{ C mol}^{-1}\times0.407 \text{ V} = -39.27 \text{ kJ mol}^{-1}$

$Cr^{2+} + 2e^- \rightarrow Cr \qquad \Delta G = 215.4 \text{ kJ mol}^{-1} - 39.27 \text{ kJ mol}^{-1} = 176.1 \text{ kJ mol}^{-1}$

$$E^\circ_{Cr^{2+}/Cr} = -\frac{\Delta G^\circ}{nF} = \frac{-176.1\times10^3 \text{J mol}^{-1}}{2\times96485 \text{ C mol}^{-1}} = -0.913 \text{ V}$$

P11.16) Determine K_{sp} for AgBr at 298.15 K using the electrochemical cell described by $\text{Ag}(s)|\text{AgBr}(s)|\text{Br}^-(aq, a_{Br^-})||\text{Ag}^+(aq, a_{Ag^+})|\text{Ag}(s)$

The half cell and overall reactions are

$AgBr + e^- \rightarrow Ag + Br^-$ $\quad E^\circ = +0.07133$ V

$Ag \rightarrow Ag^+ + e^-$ $\quad E^\circ = -0.7996$ V

$AgBr \rightarrow Ag^+ + Br^-$ $\quad E^\circ = -0.72827$ V

$$\log_{10} K_{sp} = -\frac{nE^\circ}{0.05916} = -\frac{0.729\text{ V}}{0.05916\text{ V}} = -12.310$$

$$K_{sp} = 4.90 \times 10^{-13}$$

P11.21) Consider the half-cell reaction $O_2(g) + 4H^+(aq) + 4e^- \rightarrow 2H_2O$. By what factor are n, Q, E, and E° changed if all the stoichiometric coefficients are multiplied by the factor two? Justify your answers.

n is proportional to the number of electrons transferred, and increases by the factor two. Q is squared if all stoichiometric factors are doubled. The factor by which it is increased depends on the activities of O_2 and H^+.

$E^\circ = \frac{\Delta G^\circ}{nF}$ is unchanged because both ΔG° and n are doubled.

$E = E^\circ - \frac{RT}{nF}\ln Q$ is unchanged because the squaring of Q is offset exactly by the doubling of n.

P11.26) Consider the half-cell reaction $AgCl(s) + e^- \rightleftarrows Ag(s) + Cl^-(aq)$. If $\mu^\circ(AgCl, s) = -109.71$ kJ mol^{-1}, and if $E^\circ = +0.222$ V for this half-cell, calculate the standard Gibbs energy of formation of $Cl^-(aq)$.

$$\Delta G_1^\circ = -109.71\ kJ\ mol^{-1}$$

$$\Delta G_2^\circ = (-1)\times(96{,}485\ C\ mol^{-1})\times(0.222\ V) = 21.4\ kJ\ mol^{-1}$$

$$\Delta G_3^\circ = \Delta G_f^\circ(Cl^-(aq)) = \Delta G_1^\circ + \Delta G_2^\circ = -131.1\ kJ\ mol^{-1}$$

Chapter 12: From Classical to Quantum Mechanics

P12.2) For a monatomic gas, one measure of the "average speed" of the atoms is the root mean square speed, $\mathrm{v}_{rms} = \langle \mathrm{v}^2 \rangle^{1/2} = \sqrt{\dfrac{3kT}{m}}$, in which m is the molecular mass and k is the Boltzmann constant. Using this formula, calculate the de Broglie wavelength for He and Ar atoms at 100 and at 500 K.

$$\lambda = \frac{h}{m\mathrm{v}_{\mathrm{rms}}} = \frac{h}{\sqrt{3kTm}} = \frac{6.626\times10^{-34}\,\mathrm{J\,s}}{\sqrt{3\times1.381\times10^{-23}\,\mathrm{J\,K^{-1}}\times100K\times4.003\,\mathrm{amu}\times1.661\times10^{-27}\,\mathrm{kg\,amu^{-1}}}}$$

$$= 1.26\times10^{-10}\,\mathrm{m}$$

for He at 100 K. λ=5.65×10^{-11} m for He at 500 K. For Ar, $\lambda = 4.00\times10^{-11}$ m and 1.79×10^{-11} m at 100 K and 500 K, respectively.

P12.4) Electrons have been used to determine molecular structure by diffraction. Calculate the speed of an electron for which the wavelength is equal to a typical bond length, namely, 0.150 nm.

$$\mathrm{v} = \frac{p}{m} = \frac{h}{m\lambda} = \frac{6.626\times10^{-34}\,\mathrm{J\,s}}{9.109\times10^{-31}\,\mathrm{kg}\times0.150\times10^{-9}\,\mathrm{m}} = 4.85\times10^{6}\,\mathrm{m\,s^{-1}}$$

P12.6) Pulsed lasers are powerful sources of nearly monochromatic radiation. Lasers that emit photons in a pulse of 10-ns duration with a total energy in the pulse of 0.10 J at 1000 nm are commercially available.

a) What is the average power (energy per unit time) in units of watts (1 W = 1 J/s) associated with such a pulse?
b) How many 1000-nm photons are emitted in such a pulse?

$$\text{a)}\ P = \frac{\Delta E}{\Delta t} = \frac{0.10\ \mathrm{J}}{1.0\times10^{-8}\,\mathrm{s}} = 1.0\times10^{7}\,\mathrm{J\,s^{-1}}$$

$$\text{b)}\ N = \frac{E_{pulse}}{E_{photon}} = \frac{E_{pulse}}{h\frac{c}{\lambda}} = \frac{0.10\ \mathrm{J}}{6.626\times10^{-34}\,\mathrm{J\,s^{-1}}\times\dfrac{2.998\times10^{8}\,\mathrm{m\,s^{-1}}}{1.000\times10^{6}\,\mathrm{m}}} = 5.0\times10^{17}$$

P12.8) A 1000-W gas discharge lamp emits 3.00 W of ultraviolet radiation in a narrow range centered near 280 nm. How many photons of this wavelength are emitted per second?

$$n' = \frac{E_{total}}{E_{photon}} = \frac{3.00\,\text{W} \times 1\text{J}\,\text{s}^{-1}\text{W}^{-1}}{\frac{hc}{\lambda}} = \frac{3.00\,\text{W} \times 1\text{J}\,\text{s}^{-1}\text{W}^{-1}}{\frac{6.626\times10^{-34}\,\text{J s} \times 2.998\times10^{8}\,\text{m s}^{-1}}{280\times10^{-9}\,\text{m}}} = 4.23\times10^{18}\,\text{s}^{-1}$$

P12.10) What speed does a H_2 molecule have if it has the same momentum as a photon of wavelength 280 nm?

$$p = \frac{h}{\lambda} = m_{H_2}\text{v}_{H_2}$$

$$\text{v}_{H_2} = \frac{h}{m_{H_2}\lambda} = \frac{6.626\times10^{-34}\text{J s}}{2.016\,\text{amu}\times1.661\times10^{-27}\text{kg}\left(\text{amu}\right)^{-1}\times280\times10^{-9}\text{m}} = 0.707\text{m s}^{-1}$$

P12.13) The power per unit area emitted by a blackbody is given by $P = \sigma T^4$ with $\sigma = 5.67 \times 10^{-8}\text{W m}^{-2}\,\text{K}^{-4}$. Calculate the energy radiated by a spherical blackbody of radius 0.500 m at 1000 K per second. What would the radius of a blackbody at 2500 K be if it emitted the same energy as the spherical blackbody of radius 0.500 m at 1000 K?

$$E = A\sigma T^4 = 4\pi\left(0.500\,\text{m}\right)^2 \times 5.67\times10^{-8}\text{J s}^{-1}\,\text{m}^{-2}\,\text{K}^{-4} \times \left(1000\text{ K}\right)^4 = 1.78\times10^{5}\text{J s}^{-1}$$

Because the total energy radiated by the spheres must be equal,

$$4\pi r_1^2 \sigma T_1^4 = 4\pi r_2^2 \sigma T_2^4$$

$$r_2 = \sqrt{\frac{r_1^2 T_1^4}{T_2^4}} = \sqrt{\frac{\left(0.500\,\text{m}\right)^2\left(1000\text{ K}\right)^4}{\left(2500\text{ K}\right)^4}} = 0.0800\text{ m}$$

P12.15) The power (energy per unit time) radiated by black body per unit area of surface expressed in units of W m^{-2} is given by $P = \sigma T^4$ with $\sigma = 5.67\times10^{-8}\,\text{W m}^{-2}\text{K}^{-4}$. The radius of the sun is 7.00×10^5 km and the surface temperature is 6000 K. Calculate the total energy radiated per second by the sun. Assume ideal blackbody behavior.

$$E = PA = \sigma T^4 \times 4\pi r^2$$

$$= 5.67\times10^{-8}\,\text{W m}^{-2}\text{K}^{-4} \times \left(6000\text{ K}\right)^4 \times 4\pi \times \left(7.00\times10^{8}\text{m}\right)^2$$

$$= 4.52\times10^{26}\,\text{W}$$

P12.19) If an electron passes through an electrical potential difference of 1 V, it has an energy of 1 electron-volt. What potential difference must it pass through in order to have a wavelength of 0.100 nm?

$$E = \frac{1}{2}m_e\text{v}^2 = \frac{1}{2}m_e \times \left(\frac{h}{m_e\lambda}\right)^2 = \frac{h^2}{2m_e\lambda^2}$$

$$= \frac{\left(6.626\times10^{-34}\text{J s}\right)^2}{2\times9.109\times10^{-31}\text{kg}\times\left(10^{-10}\text{m}\right)^2} = 2.41\times10^{-17}\text{J}\times\frac{1\text{eV}}{1.602\times10^{-19}\text{J}} = 150.4\text{ eV}$$

The electron must pass through an electrical potential of 150.4 V.

P12.21) The work function of platinum is 5.65 eV. What is the minimum frequency of light required to observe the photoelectric effect on Pt? If light with a 150-nm wavelength is absorbed by the surface, what is the velocity of the emitted electrons?

a) For electrons to be emitted, the photon energy must be greater than the work function of the surface.

$$E = h\nu \geq 5.65\text{ eV}\times\frac{1.602\times10^{-19}\text{J}}{\text{eV}} = 9.05\times10^{-19}\text{ J}$$

$$\nu \geq \frac{E}{h} \geq \frac{9.05\times10^{-19}\text{ J}}{6.626\times10^{-34}\text{J s}} \geq 1.37\times10^{15}\text{s}^{-1}$$

b) The outgoing electron must first surmount the barrier arising from the work function, so not all the photon energy is converted to kinetic energy.

$$E_e = h\nu - \phi = \frac{hc}{\lambda} - \phi$$

$$= \frac{6.626\times10^{-34}\text{J s}\times2.998\times10^{8}\text{m s}^{-1}}{150\times10^{-9}\text{m}} - 9.05\times10^{-19}\text{ J} = 4.19\times10^{-19}\text{ J}$$

$$\text{v} = \sqrt{\frac{2E_e}{m_e}} = \sqrt{\frac{2\times4.19\times10^{-19}\text{ J}}{9.11\times10^{-31}\text{kg}}} = 9.59\times10^{5}\text{m s}^{-1}$$

$$\Delta p_{H_2} = 6.626\times10^{-27}\text{kg m s}^{-1} = m\Delta\text{v}$$

$$\Delta\text{v}_{H_2} = \frac{6.626\times10^{-27}\text{kg m s}^{-1}}{2.016\text{ amu}\times1.661\times10^{-27}\text{ kg/amu}} = 1.98\text{ m s}^{-1}$$

$$\frac{\Delta\text{v}}{\text{v}} = \frac{1.98\text{ m s}^{-1}}{\sqrt{\frac{3RT}{M}}} = \frac{1.98\text{ m s}^{-1}}{\sqrt{\frac{3\times8.314\text{ J mol}^{-1}\text{ K}^{-1}\times298\text{ K}}{2.016\times10^{-3}\text{kg mol}^{-1}}}} = 1.03\times10^{-3}$$

Chapter 13: The Schrödinger Equation

P13.2) Consider a two-level system with $\varepsilon_1 = 3.10 \times 10^{-21}$J and $\varepsilon_2 = 6.10 \times 10^{-21}$J. If $g_2 = g_1$, what value of T is required to obtain $n_2/n_1 = 0.150$? What value of T is required to obtain $n_2/n_1 = 0.999$?

$$\frac{n_2}{n_1} = \frac{g_2}{g_1}\exp\left[\frac{-(\varepsilon_5 - \varepsilon_1)}{kT}\right]$$

$$\ln\left(\frac{n_2}{n_1}\right) = \ln\left(\frac{g_2}{g_1}\right) - \frac{(\varepsilon_5 - \varepsilon_1)}{kT}$$

$$\frac{1}{T} = \frac{k}{(\varepsilon_5 - \varepsilon_1)}\left[\ln\left(\frac{g_2}{g_1}\right) - \ln\left(\frac{n_2}{n_1}\right)\right]$$

$$T = \frac{(\varepsilon_5 - \varepsilon_1)}{k\left[\ln\left(\frac{g_2}{g_1}\right) - \ln\left(\frac{n_2}{n_1}\right)\right]}$$

$$\text{for } n_2/n_1 = 0.150 \quad T = \frac{3.00\times10^{-21}\text{J}}{1.381\times10^{-23}\text{J K}^{-1}\times\left[\ln(1) - \ln(0.150)\right]} = 115\text{K}$$

$$\text{for } n_2/n_1 = 0.999 \quad T = \frac{3.00\times10^{-21}\text{J}}{1.381\times10^{-23}\text{J K}^{-1}\times\left[\ln(1) - \ln(0.999)\right]} = 2.17\times10^{5}\text{ K}$$

P13.4) A wave traveling in the z direction is described by the wave function $\Psi(z,t) = A_1\ \mathbf{x}\sin(kz - \omega t + \phi_1) + A_2\ \mathbf{y}\sin(kz - \omega t + \phi_2)$ where $\mathbf{x}$ and $\mathbf{y}$ are vectors of unit length along the x and y axes, respectively. Because the amplitude is perpendicular to the propagation direction, $\Psi(z,t)$ represents a transverse wave.

a) What requirements must A_1 and A_2 satisfy for a plane polarized wave in the x-z plane?
b) What requirements must A_1 and A_2 satisfy for a plane polarized wave in the y-z plane?
c) What requirements must A_1 and A_2 and ϕ_1 and ϕ_2 satisfy for a plane polarized wave in a plane oriented at 45° to the x-z plane?
d) What requirements must A_1 and A_2 and ϕ_1 and ϕ_2 satisfy for a circularly polarized wave?

a) The amplitude along the x axis must oscillate, and the amplitude along the y axis must vanish. Therefore $A_1 \neq 0$ and $A_2 = 0$.

b) The amplitude along the y axis must oscillate, and the amplitude along the x axis must vanish. Therefore $A_1 = 0$ and $A_2 \neq 0$.

c) The amplitude along both the *x* and *y* axes must oscillate. Therefore $A_1 \neq 0$ and $A_2 \neq 0$. Because they must oscillate in phase, $\phi_1 = \phi_2$.

d) The amplitude along both the *x* and *y* axes must oscillate with the same amplitude. Therefore $A_1 = A_2 \neq 0$. For circularly a polarized wave, the *x* and *y* components must be out of phase by $\pi/2$. Therefore $\phi_1 = \phi_2 \pm \frac{\pi}{2}$. This can be seen by comparing the *x* and *y* amplitudes for the positive sign.

$$\Psi(z,t) = A_1\ \mathbf{x}\sin(kz - \omega t + \phi_1) + A_1\ \mathbf{y}\sin(kz - \omega t + \phi_1 + \frac{\pi}{2})$$

let $kz + \phi = kz'$

$$\Psi(z,t) = A_1\ \mathbf{x}\sin(kz' - \omega t) + A_1\ \mathbf{y}\sin(kz' - \omega t + \frac{\pi}{2})$$

$$= A_1\ \mathbf{x}\sin(kz' - \omega t) + A_1\ \mathbf{y}\left[\sin(kz' - \omega t)\cos\frac{\pi}{2} + \cos(kz' - \omega t)\sin\frac{\pi}{2}\right]$$

$$= A_1\ \mathbf{x}\sin(kz' - \omega t) + A_1\ \mathbf{y}\cos(kz' - \omega t)$$

The *x* and *y* amplitudes are $\pi/2$ out of phase and the sum of the squares of their amplitudes is a constant as required for a circle.

P13.7) Express the following complex numbers in the form $re^{i\theta}$.

a) $2 - 4i$ b) 6 c) $\frac{3+i}{4i}$ d) $\frac{8+i}{2-4i}$

In the notation $re^{i\theta}$, $r = |z| = \sqrt{a^2 + b^2}$ and $\theta = \cos^{-1}\left(\frac{\operatorname{Re} z}{|z|}\right)$.

a) $2 - 4i = 2\sqrt{5}\exp\left(i\cos^{-1}\frac{1}{\sqrt{5}}\right) = 2\sqrt{5}\exp(0.352i\pi)$

b) $6 = 6\exp\left(i\cos^{-1}1\right) = 6\exp(0)$

c) $\frac{3+i}{4i} = \frac{1}{4} - \frac{3i}{4} = \frac{\sqrt{10}}{4}\exp\left(i\cos^{-1}\frac{1}{\sqrt{10}}\right) = \frac{\sqrt{10}}{4}\exp(0.398i\pi)$

d) $\frac{8+i}{2-4i} = \frac{3}{5} + \frac{17i}{10} = \frac{\sqrt{13}}{2}\exp\left(i\cos^{-1}\frac{6}{5\sqrt{13}}\right) = \frac{\sqrt{13}}{2}\exp(0.392i\pi)$

P13.10) Determine in each of the following cases if the function in the first column is an eigenfunction of the operator in the second column. If so, what is the eigenvalue?

a) $\sin\theta\cos\phi$ $\frac{\partial}{\partial\phi}$

b) $e^{-x^2/2}$ $\frac{1}{x}\frac{d}{dx}$

c) $\sin\theta$ $\qquad \frac{\sin\theta}{\cos\theta}\frac{d}{d\theta}$

a) $\sin\theta\cos\phi$ $\qquad \frac{\partial}{\partial\phi}$

$\frac{\partial}{\partial\phi}\sin\theta\cos\phi = -\sin\theta\sin\phi.$ Not an eigenfunction

b) $e^{-\frac{1}{2}x^2}$ $\qquad \frac{1}{x}\frac{d}{dx}$

$\frac{1}{x}\frac{d}{dx}e^{-\frac{1}{2}x^2} = -e^{-\frac{1}{2}x^2}$ Eigenfunction with eigenvalue –1

c) $\sin\theta$ $\qquad \frac{\sin\theta}{\cos\theta}\frac{d}{d\theta}$

$\frac{\sin\theta}{\cos\theta}\frac{d}{d\theta}\sin\theta = \sin\theta$ Eigenfunction with eigenvalue +1

P13.12) Determine in each of the following cases if the function in the first column is an eigenfunction of the operator in the second column. If so, what is the eigenvalue?

a) $3\cos^2\theta - 1$ $\qquad \frac{1}{\sin\theta}\frac{d}{d\theta}\left(\sin\theta\frac{d}{d\theta}\right)$

b) $e^{-x^2/2}$ $\qquad \frac{d^2}{dx^2} - x^2$

c) $e^{-4i\phi}$ $\qquad \frac{d^2}{d\phi^2}$

a) $3\cos^2\theta - 1$ $\qquad \frac{1}{\sin\theta}\frac{d}{d\theta}\left(\sin\theta\frac{d}{d\theta}\right)$

$$\frac{1}{\sin\theta}\frac{d}{d\theta}\left(\sin\theta\frac{d\left(3\cos^2\theta-1\right)}{d\theta}\right) = \frac{1}{\sin\theta}\frac{d}{d\theta}\left(-6\cos\theta\sin^2\theta\right)$$

$$= \frac{1}{\sin\theta}\left(6\sin^3\theta - 12\cos^2\theta\sin\theta\right) = 6\sin^2\theta - 12\cos^2\theta$$

$$= 6 - 18\cos^2\theta = -6\left(3\cos^2\theta - 1\right)$$

Eigenfunction with eigenvalue –6.

b) $e^{-\frac{1}{2}x^2}$ $\qquad\qquad \dfrac{d^2}{dx^2}-x^2$

$$\frac{d^2 e^{-\frac{1}{2}x^2}}{dx^2}-x^2 e^{-\frac{1}{2}x^2}=-e^{-\frac{1}{2}x^2}$$

Eigenfunction with eigenvalue –1.

c) $e^{-4i\phi}$ $\qquad\qquad \dfrac{d^2}{d\phi^2}$

$$\frac{d^2 e^{-4i\phi}}{d\phi^2}=-16e^{-4i\phi}$$

Eigenfunction with eigenvalue –16.

P13.14) Which of the following wave functions are eigenfunctions of the operator *d*/*dx*? If they are eigenfunctions, what is the eigenvalue?

a) $ae^{-3x}+be^{-3ix}$ b) $\sin^2 x$ c) e^{-ix} d) $\cos ax$ e) e^{-ix^2}

a) $\dfrac{d\left(ae^{-3x}+be^{-3ix}\right)}{dx}=-3ae^{-3x}-3ibe^{-3ix}$ Not an eigenfunction

b) $\dfrac{d\sin^2 x}{dx}=2\sin x\cos x$ Not an eigenfunction

c) $\dfrac{de^{-ix}}{dx}=-ie^{-ix}$ Eigenfunction with eigenvalue –*i*

d) $\dfrac{d\cos ax}{dx}=-a\sin ax$ Not an eigenfunction

e) $\dfrac{de^{-ix^2}}{dx}=-2ixe^{-ix^2}$ Not an eigenfunction

P13.16) If two operators act on a wave function as indicated by $\hat{A}\hat{B}f(x)$, it is important to carry out the operations in succession with the first operation being that nearest to the function. Mathematically, $\hat{A}\hat{B}f(x)=\hat{A}\left(\hat{B}f(x)\right)$and $\hat{A}^2 f(x)=\hat{A}\left(\hat{A}f(x)\right)$. Evaluate the following successive operations $\hat{A}\hat{B}f(x)$. The operators $\hat{A}$ and $\hat{B}$ are listed in the first and second columns and $f(x)$is listed in the third column.

a) $\dfrac{d}{dx}$ $\qquad \dfrac{d}{dx}$ $\qquad x^2+e^{ax^2}$

b) $\dfrac{\partial^2}{\partial y^2}$ $\qquad \dfrac{\partial}{\partial x}$ $\qquad (\cos 3y)\sin^2$

c) $\frac{\partial}{\partial\theta}$ $\qquad \frac{\partial^2}{\partial\phi^2}$ $\qquad \frac{\cos\phi}{\sin\theta}$

a) $$\frac{d}{dx}\left[\frac{d\left(x^2+e^{ax^2}\right)}{dx}\right]=\frac{d}{dx}\left[2x+2axe^{ax^2}\right]=2+4a^2x^2e^{ax^2}+2ae^{ax^2}$$

b) $$\frac{\partial^2}{\partial y^2}\left[\frac{\partial\left(\cos 3y\sin^2 x\right)}{\partial x}\right]=\frac{\partial^2}{\partial y^2}[2\cos 3y\sin x\cos x]=-18\cos 3y\sin x\cos x$$

c) $$\frac{\partial}{\partial\theta}\left[\frac{\partial^2\left(\frac{\cos\phi}{\sin\theta}\right)}{\partial\phi^2}\right]=\frac{\partial}{\partial\theta}\left[-\frac{\cos\phi}{\sin\theta}\right]=\frac{\cos\phi\cos\theta}{\sin^2\theta}$$

P13.18) Find the result of operating with $\frac{d^2}{dx^2}-4x^2$ on the function e^{-ax^2}. What must the value of a be to make this function an eigenfunction of the operator?

$$\frac{d^2e^{-ax^2}}{dx^2}-4x^2e^{-ax^2}=-2ae^{-ax^2}-4x^2e^{-ax^2}+4a^2x^2e^{-ax^2}=-2ae^{-ax^2}+4\left(a^2-1\right)x^2e^{-ax^2}$$

For the function to be an eigenfunction of the operator, the terms containing $x^2e^{-ax^2}$ must vanish. This is the case if $a=\pm 1$.

P13.22) Show by carrying out the integration that $\sin(m\pi x/a)$ and $\cos(m\pi x/a)$, where m is an integer, are orthogonal over the interval $0\le x\le a$. Would you get the same result if you used the interval $0\le x\le 3a/4$? Explain your result.

$$\int_0^a\cos\left(\frac{m\pi x}{a}\right)\sin\left(\frac{m\pi x}{a}\right)dx=\left[\frac{a}{2m\pi}\sin^2\left(\frac{m\pi x}{a}\right)\right]_0^a=\frac{a}{2m\pi}\left[\sin^2(m\pi)-0\right]=0$$

$$\int_0^{\frac{3a}{4}}\cos\left(\frac{m\pi x}{a}\right)\sin\left(\frac{m\pi x}{a}\right)dx=\int_0^{\frac{3a}{4}}\cos\left(\frac{m\pi x}{a}\right)\sin\left(\frac{m\pi x}{a}\right)dx$$

$$=\left[\frac{a}{2m\pi}\sin^2\left(\frac{m\pi x}{a}\right)\right]_0^{\frac{3a}{4}}=\frac{a}{2m\pi}\left[\sin^2\left(\frac{3m\pi}{4}\right)-0\right]\neq 0$$

except for the special case $\frac{3m}{4}=n$ where n is an integer. The length of the integration interval must be n periods (for n an integer) to make the integral zero.

P13.26) Because $\int_0^d \cos\left(\frac{n\pi x}{d}\right)\cos\left(\frac{m\pi x}{d}\right)dx = 0,\ m \neq n$, the functions $\cos(n\pi x/d)$ for $n = 1, 2, 3, \ldots$ form an orthogonal set. What constant must these functions be multiplied by to form an orthonormal set?

$$1 = N^2 \int_0^d \cos\left(\frac{m\pi x}{d}\right)\cos\left(\frac{m\pi x}{d}\right)dx = N^2\left[\frac{x}{2}+\frac{d}{4m\pi}\sin\left(\frac{2m\pi x}{d}\right)\right]_0^d$$

where we have used the standard integral $\int(\cos^2 ax)\,dx = \frac{1}{2}x+\frac{1}{4a}\sin 2ax$

$$1 = N^2\left[\frac{d}{2}+\frac{d}{4m\pi}\sin(2m\pi)-\frac{0}{2}-\frac{d}{4m\pi}\sin(0)\right] = \frac{d}{2}N^2 = 1$$

$$N = \sqrt{\frac{2}{d}}$$

P13.28) Carry out the following coordinate transformations.
a) Express the point $x = 3,\ y = 2$, and $z = 1$ in spherical coordinates.
b) Express the point $r = 5,\ \theta = \frac{\pi}{4}$, and $\phi = \frac{3\pi}{4}$ in Cartesian coordinates.

a) $r = \sqrt{x^2+y^2+z^2} = \sqrt{3^2+2^2+1} = \sqrt{14}$

$$\theta = \cos^{-1}\frac{z}{\sqrt{x^2+y^2+z^2}} = \cos^{-1}\frac{1}{\sqrt{14}} = 1.30 \text{ radians}$$

$$\phi = \tan^{-1}\frac{y}{x} = \tan^{-1}\frac{2}{3} = 0.588 \text{ radians}$$

b) $x = r\sin\theta\cos\phi = 5\sin\frac{\pi}{4}\cos\frac{3\pi}{4} = -2.5$

$$y = r\sin\theta\sin\phi = 5\sin\frac{\pi}{4}\sin\frac{3\pi}{4} = 2.5$$

$$z = r\cos\theta = 5\cos\frac{\pi}{4} = \frac{5}{\sqrt{2}}$$

Chapter 15: Using Quantum Mechanics on Simple Systems

P15.2) Show that the energy eigenvalues for the free particle, $E = \frac{\hbar^2 k^2}{2m}$, are consistent with the classical result $E = \frac{1}{2}mv^2$.

$$E = \frac{1}{2}m\,\mathrm{v}^2 = \frac{p^2}{2m}$$

From the de Brogliie relation, $p = \frac{h}{\lambda}$

$E = \frac{1}{2m}\left(\frac{h}{\lambda}\right)^2 = \frac{\hbar^2 k^2}{2m}$, showing consistency between the classical and quantum result.

P15.4) Is the superposition wave function for the free particle $\psi^+(x) = A_+ e^{+i\sqrt{\frac{2mE}{\hbar^2}}x} + A\_e^{-i\sqrt{\frac{2mE}{\hbar^2}}x}$ an eigenfunction of the momentum operator? Is it an eigenfunction of the total energy operator? Explain your result.

$$-i\hbar\frac{d}{dx}\left(A_+e^{+i\sqrt{\frac{2mE}{\hbar^2}}x} + A\_e^{-i\sqrt{\frac{2mE}{\hbar^2}}x}\right) = -(i)^2\,\hbar\sqrt{\frac{2mE}{\hbar^2}}A_+e^{+i\sqrt{\frac{2mE}{\hbar^2}}x} + (i)^2\,\hbar\sqrt{\frac{2mE}{\hbar^2}}A\_e^{-i\sqrt{\frac{2mE}{\hbar^2}}x}$$

$$= \hbar k\,A_+e^{+i\sqrt{\frac{2mE}{\hbar^2}}x} - \hbar k\,A\_e^{-i\sqrt{\frac{2mE}{\hbar^2}}x}$$

This function is not an eigenfunction of the momentum operator, because the operation does not return the original function multiplied by a constant.

$$-\frac{\hbar^2}{2m}\frac{d^2}{dx^2}\left(A_+e^{+i\sqrt{\frac{2mE}{\hbar^2}}x} + A\_e^{-i\sqrt{\frac{2mE}{\hbar^2}}x}\right) = -(i)^2\frac{\hbar^2}{2m}\frac{2mE}{\hbar^2}A_+e^{+i\sqrt{\frac{2mE}{\hbar^2}}x} - (-i)^2\frac{\hbar^2}{2m}\frac{2mE}{\hbar^2}A\_e^{-i\sqrt{\frac{2mE}{\hbar^2}}x}$$

$$= E\left(A_+e^{+i\sqrt{\frac{2mE}{\hbar^2}}x} + A\_e^{-i\sqrt{\frac{2mE}{\hbar^2}}x}\right)$$

This function is an eigenfunction of the total energy operator. Because the energy is proportional to p^2, the difference in sign of the momentum of these two components does not affect the energy.

P15.6) Evaluate the normalization integral for the eigenfunctions of $\hat{H}$ for the particle in the box $\psi_n(x) = A\sin\left(\frac{n\pi x}{a}\right)$ using the trigonometric identity $\sin^2 y = \frac{1-\cos 2y}{2}$.

$$1 = \int_0^a A^2 \sin^2\left(\frac{n\pi x}{a}\right) dx$$

$$\text{let } y = \frac{n\pi x}{a};\ dx = \frac{a}{n\pi} dy$$

$$1 = A^2 \frac{a}{n\pi} \int_0^{n\pi} \sin^2 y\, dy = A^2 \frac{a}{n\pi} \int_0^{n\pi} \frac{1-\cos 2y}{2} dy = A^2 \frac{a}{n\pi}\left[\frac{y}{2} - \frac{\sin 2y}{4}\right]_0^{n\pi}$$

$$= \frac{A^2}{2}\frac{a}{n\pi} n\pi - \frac{A^2}{2}\frac{a}{n\pi}(\sin 2n\pi - \sin 0) = \frac{A^2 a}{2}$$

$$A = \sqrt{\frac{2}{a}}$$

P15.8) Calculate the probability that a particle in a one-dimensional box of length a is found between $0.31a$ and $0.35a$ when it is described by the following wave functions:

a) $\sqrt{\frac{2}{a}}\sin\left(\frac{\pi x}{a}\right)$

b) $\sqrt{\frac{2}{a}}\sin\left(\frac{3\pi x}{a}\right)$

What would you expect for a classical particle? Compare your results in the two cases with the classical result.

a)

Using the standard integral $\int \sin^2(by)\,dy = \frac{y}{2} - \frac{1}{4b}\sin(2by)$

$$P = \frac{2}{a}\int_{0.31a}^{0.35a} \sin^2\left(\frac{\pi x}{a}\right) dx = \left[\frac{x}{2} - \frac{a}{4\pi}\sin\left(\frac{2\pi x}{a}\right)\right]_{0.31a}^{0.35a}$$

$$= \frac{2}{a}\left[\frac{0.35a}{2} - \frac{a}{4\pi}\sin(0.70\pi) - \frac{0.31a}{2} + \frac{a}{4\pi}\sin(0.62\pi)\right]$$

$$= 0.04 + \frac{1}{2\pi}\left[\sin(0.62\pi) - \sin(0.70\pi)\right] = 0.059$$

b)

Using the standard integral $\int \sin^2(by)\,dy = \frac{y}{2} - \frac{1}{4b}\sin(2by)$

$$P = \frac{2}{a}\int_{0.31a}^{0.35a} \sin^2\left(\frac{3\pi x}{a}\right)dx = \frac{2}{a}\left[\frac{0.35a}{2} - \frac{a}{12\pi}\sin(2.10\pi) - \frac{0.31a}{2} + \frac{a}{12\pi}\sin(1.86\pi)\right]$$

$$= 0.04 + \frac{1}{6\pi}\left[\sin(1.86\pi) - \sin(2.10\pi)\right] = 0.0010$$

Because a classical particle is equally likely to be in any given interval, the probability will be 0.04 independent of the energy. In the ground state, the interval chosen is near the maximum of the wave function so that the quantum mechanical probability is greater than the classical probability. For the $n = 3$ state, the interval chosen is near a node of the wave function so that the quantum mechanical probability is much less than the classical probability.

P15.12) Is the superposition wave function $\psi(x) = \sqrt{\frac{2}{a}}\left[\sin\left(\frac{n\pi x}{a}\right) + \sin\left(\frac{m\pi x}{a}\right)\right]$ an eigenfunction of the total energy operator for the particle in the box?

$$\hat{H}\psi(x) = -\frac{\hbar^2}{2m}\frac{d^2}{dx^2}\left[\sqrt{\frac{2}{a}}\sin\left(\frac{n\pi x}{a}\right) + \sqrt{\frac{2}{a}}\sin\left(\frac{m\pi x}{a}\right)\right]$$

$$= \frac{h^2n^2}{8ma^2}\sqrt{\frac{2}{a}}\sin\left(\frac{n\pi x}{a}\right) + \frac{h^2m^2}{8ma^2}\sqrt{\frac{2}{a}}\sin\left(\frac{m\pi x}{a}\right)$$

Because the result is not the wave function multiplied by a constant, the superposition wave function is not an eigenfunction of the total energy operator.

P15.15) What is the solution of the time-dependent Schrödinger equation $\Psi(x,t)$ for the total energy eigenfunction $\psi_4(x) = \sqrt{\frac{2}{a}}\sin\left(\frac{4\pi x}{a}\right)$ in the particle in the box model? Write $\omega = \frac{E}{\hbar}$ explicitly in terms of the parameters of the problem.

$$\psi(x,t) = \psi(x)e^{-i\omega t} = \psi(x)e^{-i\frac{Et}{\hbar}}$$

Because $E = \frac{n^2h^2}{8ma^2} = \frac{16h^2}{8ma^2}$,

$$\psi(x,t) = \sqrt{\frac{2}{a}}\sin\left(\frac{4\pi x}{a}\right)e^{-i\frac{4\pi ht}{ma^2}}$$

P15.24) Two wave functions are distinguishable if they lead to a different probability density. Which of the following wave functions are distinguishable from $\sin\theta + i\cos\theta$?

a) $(\sin\theta + i\cos\theta)\left(\frac{\sqrt{2}}{2} + i\frac{\sqrt{2}}{2}\right)$ b) $i(\sin\theta + i\cos\theta)$

c) $-(\sin\theta + i\cos\theta)$ d) $(\sin\theta + i\cos\theta)\left(-\frac{\sqrt{2}}{2} + i\frac{\sqrt{2}}{2}\right)$

e) $(\sin\theta - i\cos\theta)$ f) $e^{i\theta}$

Two wave functions ψ_1 and ψ_2 are indistinguishable if $\psi_1^*\psi_1 = \psi_2^*\psi_2$. For the wave function in the problem, $(\sin\theta - i\cos\theta)(\sin\theta + i\cos\theta) = (\sin^2\theta + \cos^2\theta) = 1$.

a) $(\sin\theta - i\cos\theta)\left(\frac{\sqrt{2}}{2} - i\frac{\sqrt{2}}{2}\right)(\sin\theta + i\cos\theta)\left(\frac{\sqrt{2}}{2} + i\frac{\sqrt{2}}{2}\right) = (\sin^2\theta + \cos^2\theta) = 1$

indistinguishable.

b) $\left[-i(\sin\theta - i\cos\theta)\right]\left[i(\sin\theta + i\cos\theta)\right] = (\sin^2\theta + \cos^2\theta) = 1$
indistinguishable

c) $\left[-(\sin\theta - i\cos\theta)\right]\left[-(\sin\theta + i\cos\theta)\right] = (\sin^2\theta + \cos^2\theta) = 1$
indistinguishable

d) $(\sin\theta - i\cos\theta)\left(-\frac{\sqrt{2}}{2} - i\frac{\sqrt{2}}{2}\right)(\sin\theta + i\cos\theta)\left(-\frac{\sqrt{2}}{2} + i\frac{\sqrt{2}}{2}\right) = (\sin^2\theta + \cos^2\theta) = 1$

indistinguishable

e) $(\sin\theta + i\cos\theta)(\sin\theta - i\cos\theta) = (\sin^2\theta + \cos^2\theta) = 1$
indistinguishable

f) $e^{-i\theta}e^{i\theta} = 1$ indistinguishable

Chapter 17: Commuting and Noncommuting Operators and the Surprising Effects of Entanglement

P17.2) Evaluate the commutator $\left[\frac{d}{dx}-x,\frac{d}{dx}+x\right]$ by applying the operators to an arbitrary function *f*(*x*).

$$\left[\frac{d}{dx}-x,\frac{d}{dx}+x\right]f(x)=\left(\frac{d}{dx}-x\right)\left(\frac{d}{dx}+x\right)f(x)-\left(\frac{d}{dx}+x\right)\left(\frac{d}{dx}-x\right)f(x)$$

$$=\left(\frac{d}{dx}-x\right)\left(\frac{df(x)}{dx}+xf(x)\right)-\left(\frac{d}{dx}+x\right)\left(\frac{df(x)}{dx}-xf(x)\right)$$

$$=\frac{d^2f(x)}{dx^2}+f(x)+x\frac{df(x)}{dx}-x\frac{df(x)}{dx}-x^2f(x)-\frac{d^2f(x)}{dx^2}+f(x)+x\frac{df(x)}{dx}-x\frac{df(x)}{dx}+x^2f(x)$$

$=2f(x)$ Therefore,

$$\left[\frac{d}{dx}-x,\frac{d}{dx}+x\right]=2$$

P17.4) Evaluate the commutator $\left[\frac{d}{dx},x^2\right]$ by applying the operators to an arbitrary function *f*(*x*).

$$\left[\frac{d}{dx},x^2\right]f(x)=\frac{d}{dx}\left(x^2f(x)\right)-x^2\frac{d}{dx}f(x)$$

$$=2xf(x)+x^2\frac{d}{dx}f(x)-x^2\frac{d}{dx}f(x)=2xf(x)$$

$$\left[\frac{d}{dx},x^2\right]=2x$$

P17.7) Evaluate the commutator $\left[\hat{x}\frac{\partial}{\partial y},\hat{y}\right]$ by applying the operators to an arbitrary function *f*(*x*,*y*).

$$\left[\hat{x}\frac{\partial}{\partial y},\hat{y}\right]f(x,y)=x\frac{\partial yf(x,y)}{\partial y}-yx\frac{\partial f(x,y)}{\partial y}$$

$$=xf(x,y)+xy\frac{\partial f(x,y)}{\partial y}-yx\frac{\partial f(x,y)}{\partial y}=xf(x,y)$$

Therefore $\left[\hat{x}\frac{\partial}{\partial y},\hat{y}\right]=x$

P17.16) Another important uncertainty principle is encountered in time-dependent systems. It relates the lifetime of a state Δt with the measured spread in the photon energy ΔE associated with the decay of this state to a stationary state of the system. "Derive" the relation $\Delta E\,\Delta t \geq \hbar/2$ in the following steps.

a) Starting from $E = p_x^2/2m$ and $\Delta E = (dE/dp_x)\,\Delta p_x$ show that $\Delta E = \mathrm{v}_x\,\Delta p_x$.

b) Using $\mathrm{v}_x = \Delta x/\Delta t$, show that $\Delta E\,\Delta t = \Delta p_x \Delta x \geq \hbar/2$.

c) Estimate the width of a spectral line originating from the decay of a state of lifetime 1.0×10^{-9} s and 1.0×10^{-11} s in inverse seconds and inverse centimeters.

a) Starting from $E = \dfrac{p_x^2}{2m}$ and $\Delta E = \left(\dfrac{dE}{dp_x}\right)\Delta p_x$ show that $\Delta E = \mathrm{v}_x\,\Delta p_x$.

$$\frac{dE}{dp_x}\Delta p_x = \frac{p_x}{m}\Delta p_x = \mathrm{v}_x \Delta p_x$$

b) Using $\mathrm{v}_x = \dfrac{\Delta x}{\Delta t}$ show that $\Delta E\,\Delta t = \Delta p_x \Delta x \geq \dfrac{\hbar}{2}$

$$\Delta E \Delta t = \frac{\Delta x}{\Delta t}\Delta p_x \Delta t = \Delta x \Delta p_x \geq \frac{\hbar}{2}$$

c) Estimate the width of a spectral line originating from the decay of a state of lifetime 1.0×10^{-9} s and 1.0×10^{-11} s in s^{-1} and cm^{-1}.

$$\Delta E \geq \frac{\hbar}{2\Delta t} = h\Delta\nu$$

$$\Delta\nu = \frac{1}{4\pi\Delta t} = \frac{1}{4\pi\left(1.0\times 10^{-9}\mathrm{s}\right)} = 8.0\times 10^{7}\mathrm{s}^{-1}$$

$$\Delta\nu\left(\mathrm{cm}^{-1}\right) = \frac{\Delta\nu\left(\mathrm{s}^{-1}\right)}{c} = \frac{8.0\times 10^{7}\mathrm{s}^{-1}}{2.998\times 10^{10}\mathrm{cm\,s}^{-1}} = 0.00265\ \mathrm{cm}^{-1}$$

The corresponding answers for 1.0×10^{-11} s are $8.0\times 10^{9}\ \mathrm{s}^{-1}$ and $0.265\ \mathrm{cm}^{-1}$, respectively.

Chapter 18: A Quantum Mechanical Model for the Vibration and Rotation of Molecules

P18.1) The force constant for a $H^{19}F$ molecule is 966 N m^{-1}.
a) Calculate the zero point vibrational energy for this molecule for a harmonic potential.
b) Calculate the light frequency needed to excite this molecule from the ground state to the first excited state.

a)

$$E_1 = h\sqrt{\frac{k}{\mu}}\left(1+\frac{1}{2}\right) = \frac{3}{2}\times 1.055\times 10^{-34}\text{J s}\times\sqrt{\frac{966\text{ N m}^{-1}}{\frac{1.0078\times 18.9984}{1.0078+18.9984}\times 1.66\text{x}10^{-27}\text{kg amu}^{-1}}}$$

$$E_1 = 1.23\times 10^{-19}\text{J}$$

$$E_0 = \hbar\sqrt{\frac{k}{\mu}}\left(\frac{1}{2}\right) = \frac{1}{3}E_1 = 4.10\times 10^{-20}\text{J}$$

b) $$\nu = \frac{E_1 - E_0}{h} = \frac{1.23\times 10^{-19}\text{J} - 4.10\times 10^{-20}\text{J}}{6.626\times 10^{-34}\text{J s}} = 1.24\times 10^{14}\text{s}^{-1}$$

P18.3) Show by carrying out the appropriate integration that the total energy eigenfunctions for the harmonic oscillator $\psi_0(x) = \left(\frac{\alpha}{\pi}\right)^{1/4} e^{-\frac{1}{2}\alpha x^2}$ and $\psi_2(x) = \left(\frac{\alpha}{4\pi}\right)^{1/4}\left(2\alpha x^2 - 1\right)e^{-\frac{1}{2}\alpha x^2}$ are orthogonal over the interval $-\infty < x < \infty$ and that $\psi_2(x)$ is normalized over the same interval. In evaluating integrals of this type, $\int_{-\infty}^{\infty} f(x)\,dx = 0$ if *f*(*x*) is an odd function of *x* and $\int_{-\infty}^{\infty} f(x)\,dx = 2\int_{0}^{\infty} f(x)\,dx$ if *f*(*x*) is an even function of *x*.

We use the standard integrals $\int_0^{\infty} x^{2n}e^{-ax^2}dx = \frac{1\cdot 3\cdot 5\cdots(2n-1)}{2^{n+1}a^n}\sqrt{\frac{\pi}{a}}$ and

$$\int_0^{\infty} e^{-ax^2}dx = \left(\frac{\pi}{4a}\right)^{1/2}$$

$$\int_{-\infty}^{\infty}\psi_2^*(x)\psi_0(x)\,dx=\int_{-\infty}^{\infty}\left(\frac{\alpha}{4\pi}\right)^{1/4}\left(2\alpha x^2-1\right)e^{-\frac{1}{2}\alpha x^2}\left(\frac{\alpha}{\pi}\right)^{1/4}e^{-\frac{1}{2}\alpha x^2}\,dx$$

$$=\left(\frac{\alpha^2}{4\pi^2}\right)^{1/4}\int_{-\infty}^{\infty}\left(2\alpha x^2-1\right)e^{-\alpha x^2}\,dx=2\left(\frac{\alpha^2}{4\pi^2}\right)^{1/4}\int_{0}^{\infty}\left(2\alpha x^2-1\right)e^{-\alpha x^2}\,dx$$

$$=\left(\frac{\alpha^2}{4\pi^2}\right)^{1/4}\left(2\alpha\frac{1}{4\alpha}\sqrt{\frac{\pi}{\alpha}}-\frac{1}{2}\sqrt{\frac{\pi}{\alpha}}\right)=0$$

$$\int_{-\infty}^{\infty}\psi_2^*(x)\psi_2(x)\,dx=\int_{-\infty}^{\infty}\left(\frac{\alpha}{4\pi}\right)^{1/4}\left(2\alpha x^2-1\right)e^{-\frac{1}{2}\alpha x^2}\left(\frac{\alpha}{4\pi}\right)^{1/4}\left(2\alpha x^2-1\right)e^{-\frac{1}{2}\alpha x^2}\,dx$$

$$=2\left(\frac{\alpha}{4\pi}\right)^{1/2}\int_{0}^{\infty}\left(4\alpha^2x^4-4\alpha x^2+1\right)e^{-\alpha x^2}\,dx$$

$$=2\left(\frac{\alpha}{4\pi}\right)^{1/2}\left(4\alpha^2\frac{3}{2^3\alpha^2}\sqrt{\frac{\pi}{\alpha}}-4\alpha\frac{1}{2^2\alpha}\sqrt{\frac{\pi}{\alpha}}+\frac{1}{2}\sqrt{\frac{\pi}{\alpha}}\right)=2\left(\frac{\alpha}{4\pi}\right)^{1/2}\sqrt{\frac{\pi}{\alpha}}\left(\frac{3}{2}-1+\frac{1}{2}\right)=1$$

P18.7) Evaluate the average linear momentum of the quantum harmonic oscillator, $\langle p_x\rangle$, for the ground state ($n = 0$) and first two excited states ($n = 1$ and $n = 2$). Use the hint about evaluating integrals in Problem P18.3.

We use the standard integrals $\int_0^{\infty} x^{2n}e^{-ax^2}dx=\frac{1\cdot3\cdot5\cdots(2n-1)}{2^{n+1}a^n}\sqrt{\frac{\pi}{a}}$ and

$$\int_0^{\infty} e^{-ax^2}dx=\left(\frac{\pi}{4a}\right)^{1/2}$$

$$\langle p_x\rangle=\int_{-\infty}^{\infty}\psi_n^*(x)\left(-i\hbar\frac{d}{dx}\right)\psi_n\,dx$$

for $n=0$, $\langle p_x\rangle=\int_{-\infty}^{\infty}\left(\frac{\alpha}{\pi}\right)^{1/4}e^{-\frac{1}{2}\alpha x^2}\left(-i\hbar\frac{d}{dx}\right)\left(\frac{\alpha}{\pi}\right)^{1/4}e^{-\frac{1}{2}\alpha x^2}\,dx$

$$\langle p_x\rangle=\left(\frac{\alpha}{\pi}\right)^{1/2}(-i\hbar)\int_{-\infty}^{\infty}-\alpha x e^{-\alpha x^2}\,dx$$

Because the integrand is an odd function of x, $\langle p_x\rangle=0$ for $n=0$.

for $n=1$, $\langle p_x\rangle=\int_{-\infty}^{\infty}\left(\frac{4\alpha^3}{\pi}\right)^{1/4}xe^{-\frac{1}{2}\alpha x^2}\left(-i\hbar\frac{d}{dx}\right)\left(\frac{4\alpha^3}{\pi}\right)^{1/4}xe^{-\frac{1}{2}\alpha x^2}\,dx$

$$\langle p_x\rangle=\left(\frac{4\alpha^3}{\pi}\right)^{1/2}(-i\hbar)\int_{-\infty}^{\infty}x\left(1-\alpha x^2\right)e^{-\alpha x^2}\,dx$$

Because the integrand is an odd function of x, $\langle p_x\rangle=0$ for $n=1$.

for $n = 2$, $\langle p_x \rangle = \int_{-\infty}^{\infty} \left(\frac{\alpha}{4\pi}\right)^{1/4} \left(2\alpha x^2 - 1\right) e^{-\frac{1}{2}\alpha x^2} \left(-i\hbar \frac{d}{dx}\right) \left(\frac{\alpha}{4\pi}\right)^{1/4} \left(2\alpha x^2 - 1\right) e^{-\frac{1}{2}\alpha x^2} dx$

$$\langle p_x \rangle = \left(\frac{\alpha}{4\pi}\right)^{1/2} (-i\hbar) \int_{-\infty}^{\infty} \left(2\alpha x^2 - 1\right) e^{-\alpha x^2} \left(-2\alpha^2 x^3 + 4\alpha x + \alpha x\right) dx$$

$$\langle p_x \rangle = \left(\frac{\alpha}{4\pi}\right)^{1/2} (-i\hbar) \int_{-\infty}^{\infty} e^{-\alpha x^2} \left(-4\alpha^3 x^5 + 12\alpha^2 x^3 - 5\alpha x\right) dx$$

Because the integrand is an odd function of x, $\langle p_x \rangle = 0$ for $n = 3$.

The result is general. $\langle p_x \rangle = 0$ for all values of n.

P18.11) The vibrational frequency of $^1H^{35}Cl$ is 8.963 x 10^{13} s^{-1}. Calculate the force constant of the molecule. How large a mass would be required to stretch a classical spring with this force constant by 1.00 cm? Use the gravitational acceleration on Earth at sea level for this problem.

$$\nu = \frac{1}{2\pi}\sqrt{\frac{k}{\mu}};\ k = 4\pi^2 \mu \nu^2$$

$$k = 4 \times \pi^2 \times \frac{1.008 \times 34.969}{1.008 + 34.969} \text{amu} \times \frac{1.661 \times 10^{-27} \text{kg}}{\text{amu}} \times \left(8.963 \times 10^{13} \text{s}^{-1}\right)^2$$

$$k = 516 \text{ kg s}^{-2}$$

$$F = kx = mg$$

$$m = \frac{kx}{g} = \frac{516 \text{ kg s}^{-2} \times 10^{-2} \text{m}}{9.81 \text{ m s}^{-2}} = 0.525 \text{ kg}$$

P18.13) Use $\sqrt{\langle x^2 \rangle}$ as calculated in Problem P18.8 as a measure of the vibrational amplitude for a molecule. What fraction is $\sqrt{\langle x^2 \rangle}$ of the 127-pm bond length of the HCl molecule for $n = 0$, 1, and 2? The force constant for the $^1H^{35}Cl$ molecule is 516 N m^{-1}. Use your results from Problem P18.8 in this problem.

For $n = 0$,

$$\sqrt{\langle x^2 \rangle} = \left(\frac{\hbar}{2\sqrt{k\mu}}\right)^{\frac{1}{2}} = \left(\frac{1.055 \times 10^{-34} \text{J s}}{2 \times \sqrt{516 \text{ N m}^{-1} \times \frac{1.0078 \times 34.9688}{1.0078 + 34.9688} \times 1.66 \times 10^{-27} \text{kg amu}^{-1}}}\right)^{\frac{1}{2}} = 7.59 \times 10^{-12} \text{m}$$

$$\frac{\sqrt{\langle x^2 \rangle}}{\text{bond length}} = \frac{7.59 \times 10^{-12} \text{m}}{127 \times 10^{-12} \text{m}} = 5.97 \times 10^{-2}$$

For $n = 1$

$$\sqrt{\langle x^2\rangle} = \left(\frac{3\hbar}{2\sqrt{k\,\mu}}\right)^{\frac{1}{2}} = \left(\frac{3\times 1.055\times 10^{-34}\text{J s}}{2\times\sqrt{516\text{ N m}^{-1}\times\frac{1.0078\times 34.9688}{1.0078+34.9688}\times 1.66\times 10^{-27}\text{kg amu}^{-1}}}\right)^{\frac{1}{2}} = 1.31\times 10^{-11}\text{m}$$

$$\frac{\sqrt{\langle x^2\rangle}}{\text{bond length}} = \frac{1.31\times 10^{-11}\text{m}}{127\times 10^{-12}\text{m}} = 0.103$$

For $n = 2$

$$\sqrt{\langle x^2\rangle} = \left(\frac{5\hbar}{2\sqrt{k\,\mu}}\right)^{\frac{1}{2}} = \left(\frac{5\times 1.055\times 10^{-34}\text{J s}}{2\times\sqrt{516\text{ N m}^{-1}\times\frac{1.0078\times 34.9688}{1.0078+34.9688}\times 1.66\times 10^{-27}\text{kg amu}^{-1}}}\right)^{\frac{1}{2}} = 1.70\times 10^{-11}\text{m}$$

$$\frac{\sqrt{\langle x^2\rangle}}{\text{bond length}} = \frac{1.70\times 10^{-11}\text{m}}{127\times 10^{-12}\text{m}} = 0.134$$

P18.17) A gas-phase $^{1}H^{19}F$ molecule, with a bond length of 91.7 pm, rotates in a three-dimensional space.
a) Calculate the zero point energy associated with this rotation.
b) What is the smallest quantum of energy that can be absorbed by this molecule in a rotational excitation?

a) There is no zero point energy because the rotation is not constrained.

b) The smallest energy that can be absorbed is

$$E = \frac{\hbar^2}{2I}J(J+1) = \frac{\hbar^2}{2I}1(1+1) = \frac{2\times\left(1.055\,\text{x}\,10^{-34}\text{J s}\right)^2}{2\times\frac{1.0078\times 18.9984}{1.0078+18.9984}\times 1.66\times 10^{-27}\text{kg amu}^{-1}\times\left(91.7\,\text{x}\,10^{-12}\text{m}\right)^2}$$

$$E = 8.33\times 10^{-22}\text{J}$$

P18.19) The moment of inertia of $^{1}H^{35}Cl$ is 2.644×10^{-47} kg m^2. Calculate $\frac{E_{rot}}{kT}$ for $J = 0$, 5, 10, and 20 at 298 K. For which of these values of J is $\frac{E_{rot}}{kT} \approx 1$?

$$E = \frac{\hbar^2}{2I} J(J+1) = \frac{\hbar^2}{2I} 1(1+1) = \frac{\left(1.055\times10^{-34}\text{J s}\right)^2}{2\times2.664\times10^{-47}\,\text{kg m}^2} J(J+1) = 2.089\times10^{-22} J(J+1)$$

$$E_{J=0} = 0$$

$$E_{J=5} = 30\times2.089\times10^{-22}\,\text{J} = 6.267\times10^{-21}\,\text{J}$$

$$\frac{E_{J=5}}{kT} = \frac{6.267\times10^{-21}\,\text{J}}{1.381\times10^{-23}\text{J K}^{-1}\times298\text{ K}} = 1.51$$

$$E_{J=10} = 110\times2.089\times10^{-22}\,\text{J} = 2.297\times10^{-20}\,\text{J}$$

$$\frac{E_{J=10}}{kT} = \frac{2.528\times10^{-20}\,\text{J}}{1.381\times10^{-23}\text{J K}^{-1}\times298\text{ K}} = 5.55$$

$$E_{J=20} = 20\times21\times2.089\times10^{-22}\,\text{J} = 8.774\times10^{-20}\,\text{J}$$

$$\frac{E_{J=20}}{kT} = \frac{8.774\times10^{-20}\,\text{J}}{1.381\times10^{-23}\text{J K}^{-1}\times298\text{ K}} = 21.2$$

$\frac{E_{rot}}{kT} \approx 1$ for $J = 5$.

P18.20) Using the Boltzmann distribution, calculate $\frac{n_J}{n_0}$ for $^1H^{35}Cl$ for the J values of Problem P18.19 at $T = 298$ K. Does $\frac{n_J}{n_0}$ go through a maximum as J increases? If so, what can you say about the value of J corresponding to the maximum?

$$\frac{n_J}{n_0} = (2J+1)e^{-(E_J-E_0)/kT} = (2J+1)\exp\left[-E_J/kT\right]$$

$$\frac{n_0}{n_0} = 1$$

$$\frac{n_5}{n_0} = (2\times5+1)\exp\left[-\left(6.267\times10^{-21}\text{J}\right)/1.381\times10^{-23}\text{J K}^{-1}\times298\text{ K}\right] = 2.42$$

$$\frac{n_{10}}{n_0} = (2\times10+1)\exp\left[-\left(2.297\times10^{-20}\text{J}\right)/1.381\times10^{-23}\text{J K}^{-1}\times298\text{ K}\right] = 0.082$$

$$\frac{n_{20}}{n_0} = (2\times20+1)\exp\left[-\left(8.774\times10^{-20}\text{J}\right)/1.381\times10^{-23}\text{J K}^{-1}\times298\text{ K}\right] = 2.60\times10^{-8}$$

$\frac{n_J}{n_0}$ goes through a maximum because it has a value greater than one for $J = 5$. You can only conclude that $J_{max} < 10$.

P18.24) For molecular rotation, the symbol J rather than l is used as the quantum number for angular momentum. A $^1H^{35}Cl$ molecule whose bond length and force

constant are 127 pm and 516 N m^{-1}, respectively, has the rotational quantum number J = 10 and vibrational quantum number n = 0.
a) Calculate the rotational and vibrational energy of the molecule. Compare each of these energies with kT at 300 K.
b) Calculate the period for vibration and rotation. How many times does the molecule rotate during one vibrational period?

a)

$$E_{rot} = \frac{J(J+1)\hbar^2}{2\mu r^2} = \frac{10\times 11\times\left(1.055\times10^{-34}\text{J s}\right)^2}{2\times\frac{1.0078\,\text{amu}\times34.9688\,\text{amu}}{1.0078\,\text{amu}+34.9688\,\text{amu}}\times\frac{1.661\times10^{-27}\text{kg}}{\text{amu}}\times\left(1.275\times10^{-10}\text{m}\right)^2}$$

$$E_{rot} = 2.55\times10^{-20}\text{J}$$

$$\frac{E_{rot}}{kT} = \frac{2.55\times10^{-20}\text{J}}{1.381\times10^{-23}\text{J K}^{-1}\times300\text{ K}} = 6.15$$

$$E_{vib} = \left(n+\frac{1}{2}\right)\hbar\sqrt{\frac{k}{\mu}} = \frac{1}{2}\times1.055\times10^{-34}\text{J s}\times\sqrt{\frac{516\text{ N m}^{-1}}{\frac{1.0078\,\text{amu}\times34.9688\,\text{amu}}{1.0078\,\text{amu}+34.9688\,\text{amu}}\times\frac{1.661\times10^{-27}\text{kg}}{\text{amu}}}}$$

$$E_{vib} = 2.97\times10^{-20}\text{J}$$

$$\frac{E_{vib}}{kT} = \frac{2.97\times10^{-20}\text{J}}{1.381\times10^{-23}\text{J K}^{-1}\times300\text{ K}} = 7.17$$

b) Calculate the period for vibration and rotation. How many times does the molecule vibrate during one rotational period?

$$E_{rot} = \frac{1}{2}I\omega^2;\quad \omega = 2\pi\nu = \frac{2\pi}{T_{rot}}$$

$$T_{rot} = \frac{2\pi}{\omega} = \frac{2\pi}{\sqrt{\frac{2E_{rot}}{I}}}$$

$$= \frac{2\pi}{\sqrt{\frac{2\times2.55\times10^{-20}\text{J}}{\frac{1.0078\,\text{amu}\times34.9688\,\text{amu}}{1.0078\,\text{amu}+34.9688\,\text{amu}}\times\frac{1.661\times10^{-27}\text{kg}}{\text{amu}}\times\left(1.275\times10^{-10}\text{m}\right)^2}}} = 1.43\times10^{-13}\text{s}$$

$$T_{vib} = \frac{1}{\nu} = 2\pi\sqrt{\frac{\mu}{k}} = 2\pi\sqrt{\frac{\frac{1.0078\,\text{amu}\times34.9688\,\text{amu}}{1.0078\,\text{amu}+34.9688\,\text{amu}}\times\frac{1.661\times10^{-27}\text{kg}}{\text{amu}}}{516\text{ N m}^{-1}}} = 1.12\times10^{-14}\text{s}$$

It vibrates $\frac{T_{rot}}{T_{vib}} = \frac{1.43\times10^{-13}\text{s}}{1.12\times10^{-14}\text{s}} = 12.8$ times in one rotational period.

Chapter 19: The Vibrational and Rotational Spectroscopy of Diatomic Molecules

P19.1) A strong absorption band in the infrared region of the electromagnetic spectrum is observed at $\tilde{\nu} = 2170\ \text{cm}^{-1}$ for $^{12}C^{16}O$. Assuming that the harmonic potential applies, calculate the fundamental frequency ν in units of inverse seconds, the vibrational period in seconds, and the zero point energy for the molecule in joules and electron-volts.

$$\nu = \tilde{\nu}c = 2170\,\text{cm}^{-1} \times 3.00\times10^{10}\text{cm}\,\text{s}^{-1} = 6.51\times10^{13}\text{s}^{-1}$$

$$T = \frac{1}{\nu} = \frac{1}{6.51\times10^{13}\text{s}^{-1}} = 1.54\times10^{-14}\text{s}$$

$$E = \frac{1}{2}h\nu = \frac{1}{2}\times6.626\times10^{-34}\text{Js}\times6.51\times10^{13}\text{s}^{-1} = 2.16\times10^{-20}\text{J}\times\frac{6.241\times10^{18}\text{eV}}{\text{J}} = 0.134\text{eV}$$

P19.3) The force constants for H_2 and Br_2 are 575 and 246 N m^{-1}, respectively. Calculate the ratio of the vibrational state populations $\frac{n_1}{n_0}$ and $\frac{n_2}{n_0}$ at $T = 300$ K and $T = 1000$ K.

The vibrational energy is given by $E_n = \left(n+\frac{1}{2}\right)h\nu = \left(n+\frac{1}{2}\right)\hbar\sqrt{\frac{k}{\mu}}$

For H_2

$$E_0 = \frac{1}{2}\hbar\sqrt{\frac{k}{\mu}} = \frac{1}{2}\times1.055\times10^{-34}\text{J s}\times\sqrt{\frac{575\ \text{N m}^{-1}}{\frac{1.0078\times1.0078}{1.0078+1.0078}\text{amu}\times1.661\times10^{-27}\text{kg amu}^{-1}}} = 4.37\times10^{-20}\text{J}$$

$$E_1 = \frac{3}{2}\hbar\sqrt{\frac{k}{\mu}} = \frac{3}{2}\times1.055\times10^{-34}\text{J s}\times\sqrt{\frac{575\ \text{N m}^{-1}}{\frac{1.0078\times1.0078}{1.0078+1.0078}\text{amu}\times1.661\times10^{-27}\text{kg amu}^{-1}}} = 1.31\times10^{-19}\text{J}$$

$$E_2 = \frac{5}{2}\hbar\sqrt{\frac{k}{\mu}} = \frac{5}{2}\times1.055\times10^{-34}\text{J s}\times\sqrt{\frac{575\ \text{N m}^{-1}}{\frac{1.0078\times1.0078}{1.0078+1.0078}\text{amu}\times1.661\times10^{-27}\text{kg amu}^{-1}}} = 2.19\times10^{-19}\text{J}$$

For Br_2

$$E_0 = \frac{1}{2}\hbar\sqrt{\frac{k}{\mu}} = \frac{1}{2}\times1.055\times10^{-34}\text{J s}\times\sqrt{\frac{246\ \text{N m}^{-1}}{\frac{78.9183\times78.9183}{78.9183+78.9183}\text{amu}\times1.661\times10^{-27}\text{kg amu}^{-1}}} = 3.23\times10^{-21}\text{J}$$

$$E_1 = \frac{3}{2}\hbar\sqrt{\frac{k}{\mu}} = \frac{3}{2}\times 1.055\times 10^{-34}\text{J s}\times\sqrt{\frac{246\text{ N m}^{-1}}{\frac{78.9183\times 78.9183}{78.9183+78.9183}\text{amu}\times 1.661\times 10^{-27}\text{kg amu}^{-1}}} = 9.70\times 10^{-21}\text{J}$$

$$E_2 = \frac{5}{2}\hbar\sqrt{\frac{k}{\mu}} = \frac{5}{2}\times 1.055\times 10^{-34}\text{J s}\times\sqrt{\frac{246\text{ N m}^{-1}}{\frac{78.9183\times 78.9183}{78.9183+78.9183}\text{amu}\times 1.661\times 10^{-27}\text{kg amu}^{-1}}} = 1.62\times 10^{-20}\text{J}$$

For H_2 at 300 K, $\frac{n_1}{n_0} = e^{-\frac{E_1-E_0}{kT}} = e^{-\frac{(1.31-0.437)\times 10^{-19}\text{J}}{1.381\times 10^{-23}\times 300\text{ K}}} = 6.81\times 10^{-10}$

For H_2 at 1000 K, $\frac{n_1}{n_0} = e^{-\frac{E_1-E_0}{kT}} = e^{-\frac{(1.31-0.437)\times 10^{-19}\text{J}}{1.381\times 10^{-23}\times 1000\text{ K}}} = 1.78\times 10^{-3}$

For H_2 at 300 K, $\frac{n_2}{n_0} = e^{-\frac{E_1-E_0}{kT}} = e^{-\frac{(2.19-0.437)\times 10^{-19}\text{J}}{1.381\times 10^{-23}\times 300\text{ K}}} = 4.65\times 10^{-19}$

For H_2 at 1000 K, $\frac{n_2}{n_0} = e^{-\frac{E_1-E_0}{kT}} = e^{-\frac{(2.19-0.437)\times 10^{-19}\text{J}}{1.381\times 10^{-23}\times 1000\text{ K}}} = 3.07\times 10^{-6}$

For Br_2 at 300 K, $\frac{n_1}{n_0} = e^{-\frac{E_1-E_0}{kT}} = e^{-\frac{(9.70-3.23)\times 10^{-21}\text{J}}{1.381\times 10^{-23}\times 300\text{ K}}} = 0.212$

For Br_2 at 1000 K, $\frac{n_1}{n_0} = e^{-\frac{E_1-E_0}{kT}} = e^{-\frac{(9.70-3.23)\times 10^{-21}\text{J}}{1.381\times 10^{-23}\times 1000\text{ K}}} = 0.628$

For Br_2 at 300 K, $\frac{n_2}{n_0} = e^{-\frac{E_2-E_0}{kT}} = e^{-\frac{(1.62-0.323)\times 10^{-20}\text{J}}{1.381\times 10^{-23}\times 300\text{ K}}} = 4.50\times 10^{-2}$

For Br_2 at 1000 K, $\frac{n_2}{n_0} = e^{-\frac{E_2-E_0}{kT}} = e^{-\frac{(1.62-0.323)\times 10^{-20}\text{J}}{1.381\times 10^{-23}\times 1000\text{ K}}} = 0.394$

P19.6) Show that the Morse potential approaches the harmonic potential for small values of the vibrational amplitude. (*Hint:* Expand the Morse potential in a Taylor-Mclaurin series.)

$$V(R) = D_e\left[1 - e^{-\alpha(R-R_e)}\right]^2$$

Expanding in a Taylor-Mclaurin series and keeping only the first term,

$$V(R) = D_e\left[1 - \left[e^{-\alpha(R-R_e)}\right]_{R=R_e} - \left[\frac{\partial e^{-\alpha(R-R_e)}}{\partial(R-R_e)}\right]_{R=R_e}(R-R_e)\right]^2$$

$$= D_e\left[1 - 1 - \alpha(R-R_e)\right]^2 = D_e\alpha^2(R-R_e)^2$$

P19.8) The fundamental vibrational frequencies for $^1H^{19}F$ and $^2D^{19}F$ are 4138.52 and 2998.25 cm^{-1}, respectively, and D_e for both molecules is 5.86 eV. What is the difference in the bond energy of the two molecules?

$$\left(D_e - \frac{1}{2}hc\tilde{\nu}\right)_{HF} - \left(D_e - \frac{1}{2}hc\tilde{\nu}\right)_{DF} = \frac{1}{2}hc\left(\tilde{\nu}_{HF} - \tilde{\nu}_{DF}\right)$$

$$= \frac{1}{2}\times 6.626\times10^{-34}\text{J s}\times 3.00\times10^{10}\text{cm s}^{-1}\times\left(4138.52\,\text{cm}^{-1} - 2998.25\,\text{cm}^{-1}\right) = 1.133\times10^{-20}\text{J}$$

P19.14) The rotational constant for $^{127}I\,^{79}Br$ determined from microwave spectroscopy is 0.1141619 cm^{-1}. The atomic masses of ^{127}I and ^{79}Br are 126.904473 amu and 78.918336 amu, respectively. Calculate the bond length in $^{127}I^{79}Br$ to the maximum number of significant figures consistent with this information.

$$B = \frac{h}{8\pi^2\mu r_0^2};\quad r_0 = \sqrt{\frac{h}{8\pi^2\mu B}}$$

$$r_0 = \sqrt{\frac{6.6260755\times10^{-34}\text{J s}}{8\pi^2\times\frac{126.904473\times78.918336\,\text{amu}}{(126.904473+78.918336)}\times1.6605402\times10^{-27}\text{kg amu}^{-1}\times0.1141619\,\text{cm}^{-1}\times2.99792458\times10^{10}\text{cm s}^{-1}}}$$

$$r_0 = 1.742035\times10^{-10}\,\text{m}$$

P19.23) Fill in the missing step in the derivation that led to the calculation in the spectral line shape in Figure 19.22. Starting from

$$a_2(t) = \mu_x^{21}\frac{E_0}{2}\left(\frac{1 - e^{\frac{i}{\hbar}(E_2 - E_1 + h\nu)t}}{E_2 - E_1 + h\nu} + \frac{1 - e^{-\frac{i}{\hbar}(E_2 - E_1 - h\nu)t}}{E_2 - E_1 - h\nu}\right)$$ and neglecting the first term in the parenthesis, show that

$$a_2^*(t)a_2(t) = \frac{E_0}{4}\left[\mu_x^{21}\right]^2\frac{\sin^2\left[(E_2 - E_1 - h\nu)t/2\hbar\right]}{(E_2 - E_1 - h\nu)^2}$$

$$a_2^*(t)a_2(t)=\left[\mu_z^{21}\right]^2 E_0^2\left(\frac{1-e^{+\frac{i}{\hbar}(E_2-E_1-h\nu)t}}{E_2-E_1-h\nu}\right)\left(\frac{1-e^{-\frac{i}{\hbar}(E_2-E_1-h\nu)t}}{E_2-E_1-h\nu}\right)$$

$$=\left[\mu_z^{21}\right]^2 E_0^2\frac{\left(2-2\cos\left[(E_2-E_1-h\nu)\frac{t}{\hbar}\right]\right)}{(E_2-E_1-h\nu)^2}$$

Using the identity $1-\cos x=1-\cos\left(\frac{x}{2}+\frac{x}{2}\right)=\cos^2\frac{x}{2}+\sin^2\frac{x}{2}-\left(\cos^2\frac{x}{2}-\sin^2\frac{x}{2}\right)=2\sin^2\frac{x}{2}$

$$a_2^*(t)a_2(t)=\left[\mu_z^{21}\right]^2\frac{E_0^2}{4}\left(\frac{2-2\cos\left[(E_2-E_1-h\nu)\frac{t}{\hbar}\right]}{(E_2-E_1-h\nu)^2}\right)=\left[\mu_z^{21}\right]^2 E_0^2\frac{\sin^2\left[(E_2-E_1-h\nu)\frac{t}{2\hbar}\right]}{(E_2-E_1-h\nu)^2}$$

Chapter 20: The Hydrogen Atom

P20.8) Using the result of Problem P20.7, calculate the probability of finding the electron in the 1*s* state outside a sphere of radius $0.5a_0$, $3a_0$, and $5a_0$.

The probability of finding the electron inside the sphere is $1-e^{-\frac{2r}{a_0}}-\frac{2r}{a_0}\left(1+\frac{r}{a_0}\right)e^{-\frac{2r}{a_0}}$.

The probability of finding it outside the sphere is

$1-\left[1-e^{-\frac{2r}{a_0}}-\frac{2r}{a_0}\left(1+\frac{r}{a_0}\right)e^{-\frac{2r}{a_0}}\right]=e^{-\frac{2r}{a_0}}+\frac{2r}{a_0}\left(1+\frac{r}{a_0}\right)e^{-\frac{2r}{a_0}}$. Evaluating this function at $0.5a_0$, $3a_0$, and $5a_0$ gives 0.920, 0.0620, and 2.77×10^{-3}, respectively.

P20.9) The *d* orbitals have the nomenclature $d_{z^2}, d_{xy}, d_{xz}, d_{yz}$, and $d_{x^2-y^2}$. Show how the *d* orbital $\psi_{3d_{yz}}(r,\theta,\phi)=\frac{\sqrt{2}}{81\sqrt{\pi}}\left(\frac{1}{a_0}\right)^{3/2}\frac{r^2}{a_0^2}e^{-r/3a_0}\sin\theta\cos\theta\sin\phi$ can be written in the form $yzF(r)$.

In spherical coordinates, $x=r\sin\theta\cos\phi$, $y=r\sin\theta\sin\phi$, and $z=r\cos\theta$. Therefore,

$$\psi_{3d_{yz}}(r,\theta,\phi)=\frac{\sqrt{2}}{81\sqrt{\pi}}\left(\frac{1}{a_0}\right)^{3/2}\frac{r^2}{a_0^2}e^{-r/3a_0}\sin\theta\cos\theta\sin\phi$$

$$=\frac{\sqrt{2}}{81\sqrt{\pi}}\left(\frac{1}{a_0}\right)^{3/2}\frac{1}{a_0^2}e^{-r/3a_0}(r\cos\theta)(r\sin\theta\sin\phi)=\frac{\sqrt{2}}{81\sqrt{\pi}}\left(\frac{1}{a_0}\right)^{3/2}\frac{1}{a_0^2}e^{-r/3a_0}(yz)$$

P20.12) Calculate the expectation value $\left(r-\langle r\rangle\right)^2$ if the H atom wave function is $\psi_{100}(r)$.

$$\left\langle\left(r-\langle r\rangle\right)^2\right\rangle=\left\langle r^2-2r\langle r\rangle+\langle r\rangle^2\right\rangle=\langle r^2\rangle-2\langle r\rangle^2+\langle r\rangle^2=\langle r^2\rangle-\langle r\rangle^2$$

$$\langle r\rangle=\frac{1}{\pi a_0^3}\int_0^{2\pi}d\phi\int_0^{\pi}\sin\theta\,d\theta\int_0^{\infty}r^3e^{-\frac{2r}{a_0}}dr$$

$$\langle r\rangle=\frac{4}{a_0^3}\int_0^{\infty}r^3e^{-\frac{2r}{a_0}}dr$$

$\frac{4}{a_0^3}\left[\frac{3!}{\left[\frac{2}{a_0}\right]^4}\right]$ $\quad \frac{4}{a_0^3}\frac{24\,a_0^4}{16}$ $\quad$ $24a_0$

Using the standard integral $\int_0^{\infty}r^ne^{-\alpha r}=\frac{n!}{\alpha^{n+1}}$

$$\langle r\rangle = \frac{4}{a_0^3}\frac{6a_0^4}{16} = \frac{3}{2}a_0$$

$$\langle r^2\rangle = \frac{1}{\pi a_0^3}\int_0^{2\pi} d\phi \int_0^{\pi} \sin\theta\, d\theta \int_0^{\infty} r^4 e^{-\frac{2r}{a_0}} dr$$

$$\langle r^2\rangle = \frac{4}{a_0^3}\int_0^{\infty} r^4 e^{-\frac{2r}{a_0}} dr$$

Using the standard integral $\int_0^{\infty} r^n e^{-\alpha r} = \frac{n!}{\alpha^{n+1}}$

$$\langle r^2\rangle = \frac{4}{a_0^3}\frac{4!}{(2/a_0)^5} = 3(a_0)^2$$

Therefore, $\left(r-\langle r\rangle\right)^2 = \langle r^2\rangle - \langle r\rangle^2 = 3(a_0)^2 - \left(\frac{3}{2}a_0\right)^2 = \frac{3}{4}(a_0)^2$.

P20.19) The force acting between the electron and the proton in the H atom is given by $F = -\frac{e^2}{4\pi\varepsilon_0 r^2}$. Calculate the expectation value $\langle F\rangle$ for the 1*s* and $2p_z$ states of the H atom in terms of *e*, ε_0, and a_0.

$$\langle F\rangle_{1s} = -\frac{e^2}{4\pi\varepsilon_0}\int \psi^*(\tau)\frac{1}{r^2}\psi(\tau)\, d\tau$$

$$\langle F\rangle_{1s} = -\frac{e^2}{4\pi\,\varepsilon_0}\frac{1}{\pi\, a_0^3}\int_0^{2\pi} d\phi \int_0^{\pi} \sin\theta\, d\theta \int_0^{\infty}\left[e^{-r/a_0}\right]\left(\frac{1}{r^2}\right)\left[\, e^{-r/a_0}\right] r^2 d\,r$$

$$\langle F\rangle_{1s} = -\frac{e^2}{4\pi\,\varepsilon_0}\frac{4}{a_0^3}\int_0^{\infty} e^{-2r/a_0} dr = -\frac{e^2}{4\pi\,\varepsilon_0}\frac{4}{a_0^3}\left[-\frac{a_0}{2}e^{-2r/a_0}\right]_0^{\infty} = -\frac{e^2}{2\pi\,\varepsilon_0 a_0^2}$$

$$\langle F\rangle_{2pz} = -\frac{e^2}{4\pi\varepsilon_0}\int \psi^*(\tau)\frac{1}{r^2}\psi(\tau)\,d\tau$$

$$\langle F\rangle_{2pz} = -\frac{e^2}{4\pi\,\varepsilon_0}\frac{1}{32\pi\,a_0^3}\int_0^{2\pi} d\phi\int_0^{\pi}\cos^2\theta\sin\theta\,d\theta\int_0^{\infty}\left(\frac{r}{a_0}\right)^2\left[e^{-r/a_0}\right]\left(\frac{1}{r^2}\right)r^2\,d\,r$$

$$\langle F\rangle_{2pz} = -\frac{e^2}{4\pi\,\varepsilon_0}\frac{1}{16\,a_0^5}\left[\frac{\cos^3\theta}{3}\right]_0^{\pi}\times\int_0^{\infty} r^2 e^{-r/a_0}\,dr$$

$$\langle F\rangle_{2pz} = -\frac{e^2}{4\pi\,\varepsilon_0}\frac{1}{24\,a_0^3}\int_0^{\infty} r^2 e^{-r/a_0}\,dr$$

Using the standard integral $\int_0^{\infty} r^n e^{-\alpha r} = \frac{n!}{\alpha^{n+1}}$

$$\langle F\rangle_{2pz} = -\frac{e^2}{4\pi\,\varepsilon_0}\frac{1}{24\,a_0^5}\times 2a_0^3 = -\frac{e^2}{48\pi\,\varepsilon_0 a_0^2}$$

P20.22) Calculate the expectation value of the moment of inertia of the H atom in the 2*s* and $2p_z$ states in terms of μ and a_0.

For the 2*s* state,

$$\langle I\rangle = \langle \mu r^2\rangle = \mu\,\frac{1}{32\pi\,a_0^3}\int_0^{2\pi} d\phi\int_0^{\pi}\sin\theta\,d\theta\int_0^{\infty} r^4\left(2-\frac{r}{a_0}\right)^2 e^{-r/a_0}\,dr$$

$$= \mu\frac{1}{8\,a_0^3}\int_0^{\infty}\left(4r^4-\frac{4r^5}{a_0}+\frac{r^6}{a_0^2}\right)e^{-r/a_0}\,dr = \mu\frac{1}{8\,a_0^3}\left(4\int_0^{\infty} r^4\,e^{-r/a_0}\,dr-\frac{4}{a}\int_0^{\infty} r^5\,e^{-r/a_0}\,dr+\frac{1}{a^2}\int_0^{\infty} r^6\,e^{-r/a_0}\,dr\right)$$

Using the standard integral $\int_0^{\infty} r^n e^{-\alpha r} = \frac{n!}{\alpha^{n+1}}$

$$\langle I\rangle = \mu\frac{1}{8\,a_0^3}\left(4\times 4!\,a_0^5-\frac{4}{a_0}\times 5!\times a_0^6+\frac{1}{a_0^2}\times 6!\times a_0^7\right) = 42\mu a_0^2$$

For the $2p_z$ state,

$$\langle I\rangle = \langle \mu r^2\rangle = \mu\,\frac{1}{32\pi\,a_0^3}\int_0^{2\pi} d\phi\int_0^{\pi}\cos^2\theta\sin\theta\,d\theta\int_0^{\infty} r^4\left(\frac{r}{a_0}\right)^2 e^{-r/a_0}\,dr$$

$$= \mu\frac{1}{16\,a_0^5}\left[\frac{\cos^3\theta}{3}\right]_0^{\pi}\times\int_0^{\infty} r^6 e^{-r/a_0}\,dr = \mu\frac{1}{24\,a_0^5}6!\,a_0^7 = 30\mu a_0^2$$

Chapter 21: Multielectron Atoms

P21.3) Is $\psi(1,2)=1s(1)\alpha(1)1s(2)\beta(2)+1s(2)\alpha(2)1s(1)\beta(1)$ an eigenfunction of the operator $\hat{S}_z$? If so, what is its eigenvalue M_S?

$$\hat{S}_z\left[1s(1)\alpha(1)1s(2)\beta(2)+1s(2)\alpha(2)1s(1)\beta(1)\right]$$
$$=(\hat{s}_{z1}+\hat{s}_{z2})\left[1s(1)\alpha(1)1s(2)\beta(2)+1s(2)\alpha(2)1s(1)\beta(1)\right]$$
$$=\frac{\hbar}{2}\left[1s(1)\alpha(1)1s(2)\beta(2)-1s(2)\alpha(2)1s(1)\beta(1)\right]$$
$$+\frac{\hbar}{2}\left[-1s(1)\alpha(1)1s(2)\beta(2)+1s(2)\alpha(2)1s(1)\beta(1)\right]$$
$$=\frac{\hbar}{2}(0)$$

The function is an eigenfunction of $\hat{S}_z$ with the eigenvalue $M_S = 0$.

P21.4) The operator for the square of the total spin of two electrons is $\hat{S}^2_{total}=\left(\hat{S}_1+\hat{S}_2\right)^2=\hat{S}_1^2+\hat{S}_2^2+2\left(\hat{S}_{1x}\hat{S}_{2x}+\hat{S}_{1y}\hat{S}_{2y}+\hat{S}_{1z}\hat{S}_{2z}\right)$. Given that

$$\hat{S}_x\alpha=\frac{\hbar}{2}\beta \qquad \hat{S}_y\alpha=\frac{i\hbar}{2}\beta \qquad \hat{S}_z\alpha=\frac{\hbar}{2}\alpha$$
$$\hat{S}_x\beta=\frac{\hbar}{2}\alpha \qquad \hat{S}_y\beta=-\frac{i\hbar}{2}\alpha \qquad \hat{S}_z\beta=-\frac{\hbar}{2}\beta$$

show that $\alpha(1)\alpha(2)$ and $\beta(1)\beta(2)$ are eigenfunctions of the operator $\hat{S}^2_{total}$. What is the eigenvalue in each case?

$$\hat{S}^2_{total}\alpha(1)\alpha(2)$$
$$=\hat{S}_1^2\alpha(1)\alpha(2)+\hat{S}_2^2\alpha(1)\alpha(2)+2\left(\hat{S}_{1x}\hat{S}_{2x}\alpha(1)\alpha(2)+\hat{S}_{1y}\hat{S}_{2y}\alpha(1)\alpha(2)+\hat{S}_{1z}\hat{S}_{2z}\alpha(1)\alpha(2)\right)$$
$$=\alpha(2)\hat{S}_1^2\alpha(1)+\alpha(1)\hat{S}_2^2\alpha(2)+2\left(\hat{S}_{1x}\alpha(1)\hat{S}_{2x}\alpha(2)+\hat{S}_{1y}\alpha(1)\hat{S}_{2y}\alpha(2)+\hat{S}_{1z}\alpha(1)\hat{S}_{2z}\alpha(2)\right)$$
$$=\frac{3\hbar^2}{4}\alpha(1)\alpha(2)+\frac{3\hbar^2}{4}\alpha(1)\alpha(2)+2\left(\hat{S}_{1x}\alpha(1)\hat{S}_{2x}\alpha(2)+\hat{S}_{1y}\alpha(1)\hat{S}_{2y}\alpha(2)+\hat{S}_{1z}\alpha(1)\hat{S}_{2z}\alpha(2)\right)$$
$$=\frac{3\hbar^2}{4}\alpha(1)\alpha(2)+\frac{3\hbar^2}{4}\alpha(1)\alpha(2)+2\times\frac{\hbar}{2}\left(\hat{S}_{1x}\alpha(1)\beta(2)+i\hat{S}_{1y}\alpha(1)\beta(2)+\hat{S}_{1z}\alpha(1)\alpha(2)\right)$$
$$=\frac{3\hbar^2}{4}\alpha(1)\alpha(2)+\frac{3\hbar^2}{4}\alpha(1)\alpha(2)+2\times\left(\frac{\hbar}{2}\right)^2\left(\beta(1)\beta(2)+i^2\beta(1)\beta(2)+\alpha(1)\alpha(2)\right)$$
$$=\frac{3\hbar^2}{4}\alpha(1)\alpha(2)+\frac{3\hbar^2}{4}\alpha(1)\alpha(2)+\frac{2\hbar^2}{4}\alpha(1)\alpha(2)=2\hbar^2\alpha(1)\alpha(2)$$

The eigenvalue is $2\hbar^2$.

$$\hat{S}^2_{total}\beta(1)\beta(2)$$
$$= \hat{S}^2_1\beta(1)\beta(2)+\hat{S}^2_2\beta(1)\beta(2)+2\left(\hat{S}_{1x}\hat{S}_{2x}\beta(1)\beta(2)+\hat{S}_{1y}\hat{S}_{2y}\beta(1)\beta(2)+\hat{S}_{1z}\hat{S}_{2z}\beta(1)\beta(2)\right)$$
$$= \alpha(2)\hat{S}^2_1\beta(1)+\beta(1)\hat{S}^2_2\beta(2)+2\left(\hat{S}_{1x}\beta(1)\hat{S}_{2x}\beta(2)+\hat{S}_{1y}\beta(1)\hat{S}_{2y}\beta(2)+\hat{S}_{1z}\beta(1)\hat{S}_{2z}\beta(2)\right)$$
$$= \frac{3\hbar^2}{4}\beta(1)\beta(2)+\frac{3\hbar^2}{4}\beta(1)\beta(2)+2\left(\hat{S}_{1x}\beta(1)\hat{S}_{2x}\beta(2)+\hat{S}_{1y}\beta(1)\hat{S}_{2y}\beta(2)+\hat{S}_{1z}\beta(1)\hat{S}_{2z}\beta(2)\right)$$
$$= \frac{3\hbar^2}{4}\alpha(1)\alpha(2)+\frac{3\hbar^2}{4}\beta(1)\beta(2)+2\times\frac{\hbar}{2}\left(\hat{S}_{1x}\beta(1)\alpha(2)+i\hat{S}_{1y}\beta(1)\beta(2)-\hat{S}_{1z}\beta(1)\beta(2)\right)$$
$$= \frac{3\hbar^2}{4}\alpha(1)\alpha(2)+\frac{3\hbar^2}{4}\beta(1)\beta(2)+2\times\left(\frac{\hbar}{2}\right)^2\left(\alpha(1)\alpha(2)-i^2\beta(1)\beta(2)+\beta(1)\beta(2)\right)$$
$$= \frac{3\hbar^2}{4}\alpha(1)\alpha(2)+\frac{3\hbar^2}{4}\beta(1)\beta(2)+\frac{2\hbar^2}{4}\beta(1)\beta(2)=2\hbar^2\beta(1)\beta(2)$$

The eigenvalue is $2\hbar^2$.

P21.6) Write the Slater determinant for the ground-state configuration of Be.

$$\psi_{Be}=\frac{1}{\sqrt{4!}}\begin{vmatrix} 1s(1)\alpha(1) & 1s(1)\beta(1) & 2s(1)\alpha(1) & 2s(1)\beta(1) \\ 1s(2)\alpha(2) & 1s(2)\beta(2) & 2s(2)\alpha(2) & 2s(2)\beta(2) \\ 1s(3)\alpha(3) & 1s(3)\beta(3) & 2s(3)\alpha(3) & 2s(3)\beta(3) \\ 1s(4)\alpha(4) & 1s(4)\beta(4) & 2s(4)\alpha(4) & 2s(4)\beta(4) \end{vmatrix}$$

P21.8) In this problem you will prove that the ground-state energy for a system obtained using the variational method is greater than the true energy.

a) The approximate wave function Φ can be expanded in the true (but unknown) eigenfunctions ψ_n of the total energy operator in the form $\Phi=\sum_n c_n\psi_n$. Show that by substituting $\Phi=\sum_n c_n\psi_n$ in the equation $E=\frac{\int\Phi^*\hat{H}\Phi\,d\tau}{\int\Phi^*\Phi\,d\tau}$, you obtain the result

$$E=\frac{\sum_n\sum_m\int(c_n^*\psi_n^*)\hat{H}(c_m\psi_m)\,d\tau}{\sum_n\sum_m\int(c_n^*\psi_n^*)(c_m\psi_m)\,d\tau}$$

$$E=\frac{\int\Phi^*\hat{H}\Phi\,d\tau}{\int\Phi^*\Phi\,d\tau}=\frac{\int\left(\Phi=\sum_n c_n\psi_n\right)^*\hat{H}\left(\Phi=\sum_m c_m\psi_m\right)d\tau}{\int\left(\Phi=\sum_n c_n\psi_n\right)^*\left(\Phi=\sum_m c_m\psi_m\right)d\tau}=\frac{\sum_n\sum_m\int(c_n^*\psi_n^*)\hat{H}(c_m\psi_m)\,d\tau}{\sum_n\sum_m\int(c_n^*\psi_n^*)(c_m\psi_m)\,d\tau}$$

b) Because the ψ_n are eigenfunctions of $\hat{H}$, they are orthogonal and $\hat{H}\psi_n = E_n\psi_n$. Show that this information allows us to simplify the previous expression for E from part (a) to $E = \frac{\sum_m E_m c_m^* c_m}{\sum_m c_m^* c_m}$

$$E = \frac{\sum_n \sum_m \int (c_n^* \psi_n^*) \hat{H} (c_m \psi_m) d\tau}{\sum_n \sum_m \int (c_n^* \psi_n^*)(c_m \psi_m) d\tau} = \frac{\sum_n \sum_m E_m \int (c_n^* \psi_n^*)(c_m \psi_m) d\tau}{\sum_n \sum_m \int (c_n^* \psi_n^*)(c_m \psi_m) d\tau}$$

$$= \frac{\sum_m E_m c_m^* c_m \int \psi_m^* \psi_m d\tau}{\sum_m c_m^* c_m \int \psi_m^* \psi_m d\tau} = \frac{\sum_m E_m c_m^* c_m}{\sum_m c_m^* c_m}$$

c) Arrange the terms in the summation such that the first energy is the true ground-state energy E_0 and the energy increases with the summation index m. Why can you conclude that $E - E_0 \geq 0$?

$$E - E_0 = \frac{\sum_m E_m c_m^* c_m}{\sum_m c_m^* c_m} - \frac{E_0 \sum_m c_m^* c_m}{\sum_m c_m^* c_m} = \frac{\sum_m (E_m - E_0) c_m^* c_m}{\sum_m c_m^* c_m} \geq 0.$$ Both $(E_m - E_0)$ and $c_m^* c_m$ are greater than zero. Therefore, $E - E_0 \geq 0$.

P21.12) List the quantum numbers L and S that are consistent with the terms
a) ^{4}S, b) ^{4}G, c) ^{3}P, and d) ^{2}D

a) ^{4}S: $L = 0$, $2S + 1 = 4$, $S = 3/2$
b) ^{4}G: $L = 4$, $2S + 1 = 4$, $S = 3/2$
c) ^{3}P: $L = 1$, $2S + 1 = 3$, $S = 1$
d) ^{2}D: $L = 2$, $2S + 1 = 2$, $S = ½$

P21.14) Two angular momenta with quantum numbers $j_1 = 3/2$ and $j_2 = 5/2$ are added together. What are the possible values of J for the resultant angular momentum states?

$J = |J_1 + J_2|, |J_1 + J_2 - 1|, |J_1 + J_2 - 2|, \ldots, |J_1 - J_2|$ giving possible J values of 4, 3, 2, and 1.

P21.16) A general way to calculate the number of states that arise from a given configuration is as follows. Calculate the combinations of m_l and m_s for the first electron, and call that number n. The number of combinations used is
the number of electrons which we call m. The number of unused combinations is $n - m$. According to probability theory, the number of distinct permutations that arise from

distributing the m electrons among the n combinations is $\dfrac{n!}{m!(n-m)!}$. For example, the number of states arising from a p^2 configuration is $\dfrac{6!}{2!4!}=15$, which is the result obtained in Section 21.8. Using this formula, calculate the number of possible ways to place five electrons in a d subshell. What is the ground-state term for the d^5 configuration and how many states does the term include?

The first electrons can have any combination of 5 m_l and 2 m_s values so that $n = 10$ and $m = 5$. Using the formula, the calculated number of states is $\dfrac{10!}{5!(10-5)!}=252$. The number of states in a term is (2L+1)(2S+1). The terms from Table 21.5 are listed here, with the number of states in each term in square brackets following the term designation: ^{6}S[6×1], ^{4}G[4×9], ^{4}F[4×7], ^{4}D[4×5], ^{4}P[4×3], ^{2}I[2×13], ^{2}H[2×11], ^{2}G(2)[$2\times2\times9$], ^{2}F(2)[$2\times2\times7$], ^{2}D(3)[$3\times2\times5$], ^{2}P[2×3], and ^{2}S[2×1]. The number of states in these terms is 6 + 36 + 28 + 20 + 12 + 26 + 22 + 36 + 28 + 30 + 6 + 2 = 252, showing consistency.

The ground state is that with the maximum multiplicity which according to Table 21.5 is the ^{6}S term. The number of states in the term = $(2L + 1)(2S + 1) = 6$.

P21.18) What atomic terms are possible for the following electron configurations? Which of the possible terms has the lowest energy?

a) ns^1np^1 b) ns^1nd^1 c) ns^2np^1 d) ns^1np^2

a) ns^1np^1 L can only have the value 1, and S can have the values 0 and 1. The possible terms are ^{1}P and ^{3}P. Hund's Rules predict that the ^{3}P term will have the lower energy.

b) ns^1nd^1 L can only have the value 2, and S can have the values 0 and 1. The possible terms are ^{1}D and ^{3}D. Hund's Rules predict that the ^{3}D term will have the lower energy.

c) ns^2np^1 L can only have the value 1, and S can only have the value 1/2. The only possible term is ^{2}P.

d) ns^1np^2 A table such as the table in the text for the p^2 configuration will have three columns, one for each of the electrons, for M_L and M_S. Each of the fifteen states for the p^2 configuration can be combined with $m_s=\pm\dfrac{1}{2}$ for the ns electron. This gives a total of 30 states. Working through the table gives ^{2}D, ^{4}P, ^{2}P, and ^{2}S terms. Hund's Rules predict that the ^{4}P term will have the lowest energy.

P21.20) Derive the ground-state term symbols for the following configurations: a) d^2, b) f^9, c) f^{14}

The method illustrated in Example Problem 21.7 is used for all parts.

a)

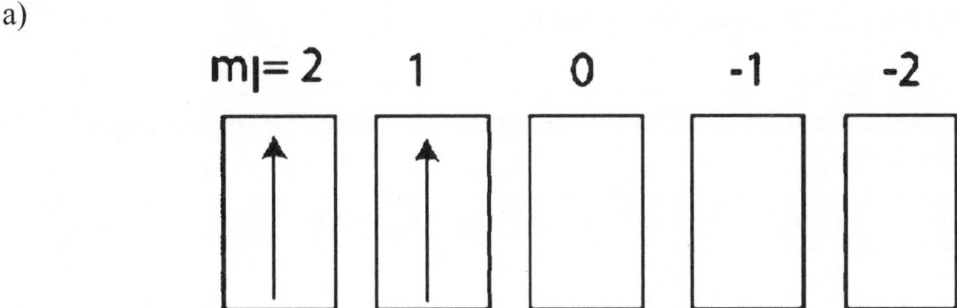

$M_{Lmax} = 3$ and $M_{Smax} = 1$. Therefore, the ground state term is ^{3}F.

b)

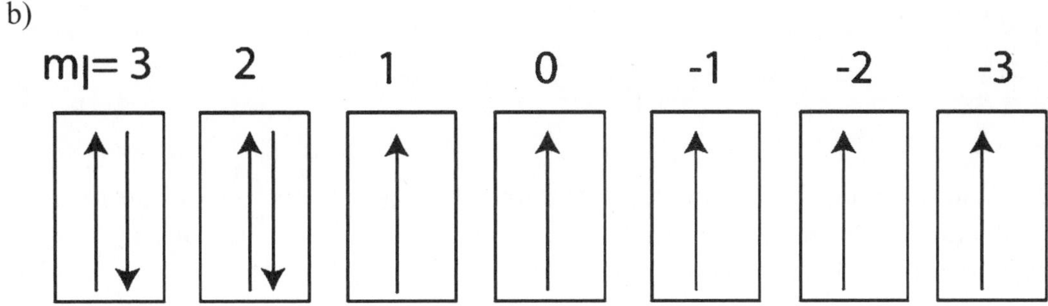

$M_{Lmax} = 5$ and $M_{Smax} = 2.5$. Therefore, the ground state term is ^{6}H.

c)

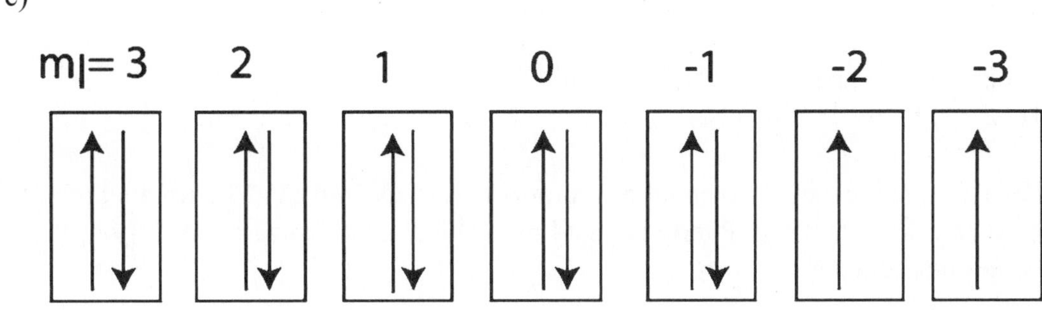

$M_{Lmax} = 5$ and $M_{Smax} = 1$. Therefore, the ground state term is ^{3}H.

P21.24) Using Table 21.5, which lists the possible terms that arise from a given configuration, and Hund's rules, write the term symbols for the ground state of the atoms K through Cu, excluding Cr, in the form $^{(2S+1)}L_J$.

We use Hund's rule that the term with the highest multiplicity is the lowest in energy to get the left superscript. For the right subscript, if there are several choices for J, the lowest J value gives the lowest energy if the subshell is less than half full, and the highest

J value gives the lowest energy if the subshell is exactly or more than half full. Applying these rules gives rise to the following term symbols.

Atom	Configuration	Ground state term symbol
K	$4s^1$	$^2S_{1/2}$
Ca	$4s^2$	$^1S_0$
Sc	$3d^1$	$^2D_{3/2}$
Ti	$3d^2$	$^3F_2$
V	$3d^3$	$^4F_{3/2}$
Mn	$3d^5$	$^6S_{5/2}$
Fe	$3d^6$	$^5D_4$
Co	$3d^7$	$^4F_{9/2}$
Ni	$3d^8$	$^3F_4$
Cu	$4s^1 3d^{10}$	$^2S_{1/2}$

P21.27) What J values are possible for a ^{6}H term? Calculate the number of states associated with each level and show that the total number of states is the same as calculated from $(2S + 1)(2L + 1)$.

$S = 5/2, L = 5.$

J lies between $|L+S|$ and $|L-S|$ and can have the values 15/2, 13/2, 11/2, 9/2, 7/2, and 5/2. The number of states is $2J + 1$ or 16, 14, 12, 10, 8, and 6, respectively.

This gives a total number of states of $16 + 14 + 12 + 10 + 8 + 6 = 66$.
$(2S + 1)(2L + 1) = 6\times 11 = 66$ also.

P21.29) The first ionization potential of ground-state He is 24.6 eV. The wavelength of light associated with the $1s2p$ ^{1}P term is 58.44 nm. What is the ionization energy of the He atom in this excited state?

The photon energy is

$$E = \frac{hc}{\lambda} = \frac{6.626\times10^{-34}\,\text{J s}\times2.998\times10^{8}\,\text{m s}^{-1}}{58.44\times10^{-9}\,\text{m}}\times\frac{1\text{J}}{1.602\times10^{-19}\,\text{J}} = 21.2\text{eV}$$

Therefore, the ionization energy of the He atom in this state is 24.6eV – 21.2eV = 3.4eV.

Chapter 22: Examples of Spectroscopy Involving Atoms

P22.1) Calculate the wavelengths of the first three lines of the Lyman, Balmer, and Paschen series, and the series limit (the shortest wavelength) for each series.

Lyman Series: $E_n = R_H\left(\frac{1}{1^2}-\frac{1}{n^2}\right)$

$$n=2 \quad E_2 = R_H\left(1-\frac{1}{4}\right)=\frac{3}{4}R_H = 82258 \text{ cm}^{-1} \quad \lambda = 121.6 \text{ nm}$$

$$n=3 \quad E_2 = R_H\left(1-\frac{1}{9}\right)=\frac{8}{9}R_H = 97491 \text{ cm}^{-1} \quad \lambda = 102.6 \text{ nm}$$

$$n=4 \quad E_2 = R_H\left(1-\frac{1}{16}\right)=\frac{15}{16}R_H = 102823 \text{ cm}^{-1} \quad \lambda = 97.3 \text{ nm}$$

$$n=\infty \quad E_2 = R_H\left(1-\frac{1}{\infty}\right)=R_H = 109677 \text{ cm}^{-1} \quad \lambda = 91.2 \text{ nm}$$

Balmer Series: $E_n = R_H\left(\frac{1}{2^2}-\frac{1}{n^2}\right)$

$$n=3 \quad E_2 = R_H\left(\frac{1}{4}-\frac{1}{9}\right)=\frac{5}{36}R_H = 15233 \text{ cm}^{-1} \quad \lambda = 656.5 \text{ nm}$$

$$n=3 \quad E_2 = R_H\left(\frac{1}{4}-\frac{1}{16}\right)=\frac{3}{16}R_H = 20565 \text{ cm}^{-1} \quad \lambda = 486.3 \text{ nm}$$

$$n=4 \quad E_2 = R_H\left(\frac{1}{4}-\frac{1}{25}\right)=\frac{21}{100}R_H = 23032 \text{ cm}^{-1} \quad \lambda = 434.2 \text{ nm}$$

$$n=\infty \quad E_2 = R_H\left(\frac{1}{4}-\frac{1}{\infty}\right)=\frac{1}{4}R_H = 27419 \text{ cm}^{-1} \quad \lambda = 364.7 \text{ nm}$$

Paschen Series: $E_n = R_H\left(\frac{1}{3^2}-\frac{1}{n^2}\right)$

$$n=4 \quad E_2 = R_H\left(\frac{1}{9}-\frac{1}{16}\right)=\frac{7}{144}R_H = 5331.5 \text{ cm}^{-1} \quad \lambda = 1876 \text{ nm}$$

$$n=5 \quad E_2 = R_H\left(\frac{1}{9}-\frac{1}{25}\right)=\frac{16}{225}R_H = 7799.3 \text{ cm}^{-1} \quad \lambda = 1282 \text{ nm}$$

$$n=6 \quad E_2 = R_H\left(\frac{1}{9}-\frac{1}{36}\right)=\frac{1}{12}R_H = 9139.8 \text{ cm}^{-1} \quad \lambda = 1094 \text{ nm}$$

$$n=\infty \quad E_2 = R_H\left(\frac{1}{9}-\frac{1}{\infty}\right)=\frac{1}{9}R_H = 12186.4 \text{ cm}^{-1} \quad \lambda = 820.6 \text{ nm}$$

P22.4) The spectrum of the hydrogen atom reflects the splitting of the $1s^2$ S and $2p^2$ P terms into levels. The energy difference between the levels in each term is much smaller than the difference in energy between the terms. Given this information, how many spectral lines are observed in the $1^2\mathrm{S} \rightarrow 2^2\mathrm{P}$ transition? Are the frequencies of these transitions very similar or quite different?

The $^2\mathrm{S}$ term has a single level, $^2\mathrm{S}_{1/2}$. The $^2\mathrm{P}$ term splits into two levels, $^2\mathrm{P}_{1/2}$ and $^2\mathrm{P}_{3/2}$. Therefore, there will be two closely spaced lines in the spectrum corresponding to the transitions $^2\mathrm{S}_{1/2}\rightarrow{}^2\mathrm{P}_{1/2}$ and $^2\mathrm{S}_{1/2}\rightarrow{}^2\mathrm{P}_{3/2}$. The energy spacing between the lines will be much smaller than the energy of the transition.

P22.6) Use the transition frequencies shown in Example Problem 22.2 to calculate the energy (in joules and electron-volts) of the six levels relative to the 3s $^2\mathrm{S}_{1/2}$ level. State your answers with the correct number of significant figures.

$$E\left(3p\,^2\mathrm{P}_{1/2}\right)=\frac{hc}{\lambda}=\frac{6.626\times10^{-34}\mathrm{J\,s}\times2.998\times10^{8}\mathrm{m\,s^{-1}}}{589.6\times10^{-9}\mathrm{m}}=3.369\times10^{-19}\mathrm{J}=2.102\ \mathrm{eV}$$

$$E\left(3p\,^2\mathrm{P}_{3/2}\right)=\frac{hc}{\lambda}=\frac{6.626\times10^{-34}\mathrm{J\,s}\times2.998\times10^{8}\mathrm{m\,s^{-1}}}{589.0\times10^{-9}\mathrm{m}}=3.373\times10^{-19}\mathrm{J}=2.105\ \mathrm{eV}$$

$$E\left(4s\,^2\mathrm{S}_{1/2}\right)=\frac{hc}{\lambda}=\frac{6.626\times10^{-34}\mathrm{J\,s}\times2.998\times10^{8}\mathrm{m\,s^{-1}}}{589.6\times10^{-9}\mathrm{m}}+\frac{6.626\times10^{-34}\mathrm{J\,s}\times2.998\times10^{8}\mathrm{m\,s^{-1}}}{1183.3\times10^{-9}\mathrm{m}}$$

$$=5.048\times10^{-19}\mathrm{J}=3.150\ \mathrm{eV}$$

$$E\left(5s\,^2\mathrm{S}_{1/2}\right)=\frac{hc}{\lambda}=\frac{6.626\times10^{-34}\mathrm{J\,s}\times2.998\times10^{8}\mathrm{m\,s^{-1}}}{589.0\times10^{-9}\mathrm{m}}+\frac{6.626\times10^{-34}\mathrm{J\,s}\times2.998\times10^{8}\mathrm{m\,s^{-1}}}{616.0\times10^{-9}\mathrm{m}}$$

$$=6.597\times10^{-19}\mathrm{J}=4.118\ \mathrm{eV}$$

$$E\left(3d\,^2\mathrm{D}_{3/2}\right)=\frac{hc}{\lambda}=\frac{6.626\times10^{-34}\mathrm{J\,s}\times2.998\times10^{8}\mathrm{m\,s^{-1}}}{589.6\times10^{-9}\mathrm{m}}+\frac{6.626\times10^{-34}\mathrm{J\,s}\times2.998\times10^{8}\mathrm{m\,s^{-1}}}{818.3\times10^{-9}\mathrm{m}}$$

$$=5.797\times10^{-19}\mathrm{J}=3.618\ \mathrm{eV}$$

$$E\left(4d\,^2\mathrm{D}_{3/2}\right)=\frac{hc}{\lambda}=\frac{6.626\times10^{-34}\mathrm{J\,s}\times2.998\times10^{8}\mathrm{m\,s^{-1}}}{589.0\times10^{-9}\mathrm{m}}+\frac{6.626\times10^{-34}\mathrm{J\,s}\times2.998\times10^{8}\mathrm{m\,s^{-1}}}{568.2\times10^{-9}\mathrm{m}}$$

$$=6.869\times10^{-19}\mathrm{J}=4.287\ \mathrm{eV}$$

P22.10) In the Na absorption spectrum, the following transitions are observed:

$4p\,^2\mathrm{P}-3s\,^2\mathrm{S}$ $\lambda=330.26$ nm
$3p\,^2\mathrm{P}-3s\,^2\mathrm{S}$ $\lambda=589.593$ nm, 588.996 nm
$5s\,^2\mathrm{S}-3p\,^2\mathrm{P}$ $\lambda=616.073$ nm, 615.421 nm

Calculate the energies of the $4p\,^2\mathrm{P}$ and $5s\,^2\mathrm{S}$ states with respect to the $3s\,^2\mathrm{S}$ ground state.

$$E\left(4p\,^2\mathrm{P}\right)=\frac{hc}{\lambda}=\frac{6.626\times10^{-34}\mathrm{J\,s}\times2.998\times10^{8}\mathrm{m\,s^{-1}}}{330.26\times10^{-9}\mathrm{m}}=6.015\times10^{-19}\mathrm{J}=3.754\ \mathrm{eV}$$

By looking at the Grotrian diagram of Example Problem 22.2, it is seen that the $5s\,^2\mathrm{S}$ state is accessed by absorption of the photons of wavelength 588.996 nm and 616.073 nm.

$$E\left(5s\,^2S\right) = \frac{hc}{\lambda} = \frac{6.626\times10^{-34}\text{J s}\times2.998\times10^{8}\text{m s}^{-1}}{5888.996\times10^{-9}\text{m}} + \frac{6.626\times10^{-34}\text{J s}\times2.998\times10^{8}\text{m s}^{-1}}{616.073\times10^{-9}\text{m}}$$

$$= 6.597\times10^{-19}\text{J} = 4.117\text{ eV}$$

P22.16) The inelastic mean free path of electrons in a solid, λ, governs the surface sensitivity of techniques such as AES and XPS. The electrons generated below the surface must make their way to the surface without losing energy in order to give elemental and chemical shift information. An empirical expression for elements that gives λ as a function of the kinetic energy of the electron generated in AES or XPS is $\lambda = 538E^{-2} + 0.41(lE)^{0.5}$. The units of λ are monolayers, E is the kinetic energy of the electron, and l is the monolayer thickness in nanometers. On the basis of this equation, what kinetic energy maximizes the surface sensitivity for a monolayer thickness of 0.3 nm? An equation solver would be helpful in obtaining the answer.

$$\frac{d}{dE}\left(538E^{-2} + 0.41\left(lE\right)^{0.5}\right) = -\frac{1076}{E^3} + \frac{0.1123}{\sqrt{E}}$$

Setting the derivative equal to zero and solving for E gives two complex roots and $E = 39$ eV.

Chapter 30: Probability

P30.1) Suppose that you draw a card from a standard deck of 52 cards. What is the probability of drawing:
a) an ace of any suit?
b) the ace of spades?
c) How would your answers to parts (a) and (b) change if you were allowed to draw three times, replacing the card drawn back into the deck after each draw?

a) In a deck of 52 cards there are four aces, therefore:

$$P_E = \frac{E}{N} = \frac{4}{52}$$

b) There is only one card that corresponds to the event of interest, therefore:

$$P_E = \frac{E}{N} = \frac{1}{52}$$

c) By replacing the card, the probability for each drawing is independent, and the total probability is the sum of probabilities for each event. Therefore, the probability will be $3P_E$ with P_E as given above.

P30.3) A pair of standard dice are rolled. What is the probability of observing the following:
a) The sum of the dice is equal to 7.
b) The sum of the dice is equal to 9.
c) The sum of the dice is less than or equal to 7.

a) We are interested in the outcome where the sum of two dice is equal to 7. If any side of a die has an equal probability of being observed, then the probability of any number appearing is 1/6.

$$\begin{aligned} P_{sum=7} &= \left[2\times(P_1\times P_6)\right]+\left[2\times(P_2\times P_5)\right]+\left[2\times(P_3\times P_4)\right] \\ &= 3\left[\frac{2}{36}\right] = \frac{1}{6} \end{aligned}$$

b) Using the nomenclature developed above:

$$P_{sum=9} = [2\times(P_3\times P_6)]+[2\times(P_4\times P_5)]$$
$$= 2\left[\frac{2}{36}\right] = \frac{1}{18}$$

c) Now, one has to sum all the probabilities that correspond to the event of interest:

$$P_{sum<7} = [(P_1\times P_1)]+[2\times(P_1\times P_2)]+[2\times(P_1\times P_3)]+[2\times(P_1\times P_4)]+[2\times(P_1\times P_5)]+[2\times(P_1\times P_6)]$$
$$+[(P_2\times P_2)]+[2\times(P_2\times P_3)]+[2\times(P_2\times P_4)]+[2\times(P_2\times P_5)]+[(P_3\times P_3)]+[2\times(P_3\times P_4)]$$
$$= 21\left[\frac{1}{36}\right] = \frac{21}{36}$$

P30.7) Determine the numerical values for the following:
a) The number of configurations employing all objects in a six-object set
b) The number of configurations employing 4 objects from a six-object set
c) The number of configurations employing no objects from a six-object set
d) $C(50,10)$

a) $C(n,j)=C(6,6)=\left(\frac{n!}{j!(n-j)!}\right)=\frac{6!}{6!0!}=1$

b) $C(n,j)=C(6,4)=\left(\frac{n!}{j!(n-j)!}\right)=\frac{6!}{4!2!}=15$

c) $C(n,j)=C(6,0)=\left(\frac{n!}{j!(n-j)!}\right)=\frac{6!}{0!6!}=1$

d) $P(n,j)=P(50,10)=\left(\frac{n!}{j!(n-j)!}\right)=\frac{50!}{10!40!}\cong 1.03\times10^{10}$

P30.9) Four bases (A, C, T, and G) appear in DNA. Assume that the appearance of each base in a DNA sequence is random.
a) What is the probability of observing the sequence AAGACATGCA?
b) What is the probability of finding the sequence GGGGGAAAAA?
c) How do your answers to parts (a) and (b) change if the probability of observing A is twice that of the probabilities used in parts (a) and (b) of this question when the preceding base is G?

a) There are four choices for each base, and the probability of observing any base is equal. Therefore, for a decamer the number of possible sequences is:

$$N_{total} = (4)^{10} \cong 1.05\times10^{6}$$

Since there is only one sequence that corresponds to the event of interest:

$$P_E = \frac{E}{N} \cong \frac{1}{1.03\times 10^6} \cong 9.54\times 10^{-7}$$

b) Identical to part (a).

c) In this case, the probability of observing a base at a given location is dependent on which base is present. If G appears in the sequence, then the probability of observing A is 1/2 while the probability of observing any other base is 1/6 (watch the normalization!). Therefore, the probability of observing the sequence in part (a) is:

$$P = \left(\frac{1}{4}\right)^3\left(\frac{1}{2}\right)\left(\frac{1}{4}\right)^4\left(\frac{1}{6}\right)\left(\frac{1}{4}\right) \cong 1.27\times 10^{-6}$$

and the probability for the sequence in part (b) is:

$$P = \left(\frac{1}{4}\right)\left(\frac{1}{6}\right)^4\left(\frac{1}{2}\right)\left(\frac{1}{4}\right)^4 \cong 3.77\times 10^{-7}$$

P30.11) The Washington State Lottery consists of drawing five balls numbered 1 to 43, and a single ball numbered 1 to 23 from a separate machine.
a) What is the probability of hitting the jackpot in which the values for all six balls are correctly predicted?
b) What is the probability of predicting just the first five balls correctly?
c) What is the probability of predicting the first five balls in the exact order they are picked?

a) The total probability is the product of probabilities for the five-ball outcome and the one-ball outcome. The five-ball outcome is derived by considering the configurations possible using five objects from a set of 43 total objects:

$$P_{fiveball} = \left[C(43,5)\right]^{-1} = \left(\frac{43!}{5!38!}\right)^{-1} \cong 1.04\times 10^{-6}$$

The one-ball outcome is associated with the configurations possible using a single object from a set of 23 objects:

$$P_{oneball} = \left[C(23,1)\right]^{-1} = \left(\frac{23!}{1!22!}\right)^{-1} \cong 4.35\times 10^{-2}$$

The total probability is the product of the above probabilities:

$$P_{total} = P_{fiveball}\times P_{oneball} \cong 4.52\times 10^{-8}$$

b) The probability is that for the five-ball case determined above.

c) This case corresponds to a specific permutation of all permutations possible using five objects from a set of 43 objects:

$$P = [P(43,5)]^{-1} = \left(\frac{43!}{38!}\right)^{-1} \cong 8.66\times10^{-9}$$

P30.14) Imagine an experiment in which you flip a coin four times. Furthermore, the coin is balanced fairly such that the probability of landing heads or tails is equivalent. After tossing the coin 10 times, what is the probability of observing

a) no heads?
b) two heads?
c) five heads?
d) eight heads?

a) The quantity of interest is the probability of observing a given number of successful trials (j) in a series of n trials in which the probability of observing a successful trial, P_E, is equal to 1/2:

$$P(j) = C(n,j)(P_E)^j(1-P_E)^{n-j} = C(n,j)\left(\frac{1}{2}\right)^n$$

Substituting in for the specific case of $j = 0$ and $n = 10$ yields:

$$P(0) = C(10,0)\left(\frac{1}{2}\right)^{10} \cong 9.77\times10^{-4}$$

b) In this case, $j = 2$ and $n = 10$:

$$P(2) = C(10,2)\left(\frac{1}{2}\right)^{10} = \left(\frac{10!}{2!8!}\right)\left(\frac{1}{2}\right)^{10} \cong 0.044$$

c) In this case, $j = 5$ and $n = 10$:

$$P(5) = C(10,5)\left(\frac{1}{2}\right)^{10} = \left(\frac{10!}{5!5!}\right)\left(\frac{1}{2}\right)^{10} \cong 0.246$$

d) In this case, $j = 8$ and $n = 10$:

$$P(8) = C(10,8)\left(\frac{1}{2}\right)^{10} = \left(\frac{10!}{8!2!}\right)\left(\frac{1}{2}\right)^{10} \cong 0.044$$

P30.19) Radioactive decay can be thought of as an exercise in probability theory. Imagine that you have a collection of radioactive nuclei at some initial time (N_0) and are interested in how many nuclei will still remain at a later time (N). For first-order radioactive decay, $N/N_0 = e^{-kt}$. In this expression, k is known as the decay constant and t is time.

a) What is the variable of interest in describing the probability distribution?
b) At what time will the probability of nuclei undergoing radioactive decay be 0.50

a) The variable (k) defines the width of the distribution of population versus time.

b)

$$\frac{N}{N_0} = 0.5 = e^{-kt}$$
$$\ln(0.5) = -kt$$
$$\frac{-\ln(0.5)}{t} = \frac{\ln(2)}{t} = k$$

P30.20) In Chapter 31, we will encounter the energy distribution $P(\varepsilon) = Ae^{-\varepsilon/kT}$, where $P(\varepsilon)$ is the probability of a molecule occupying a given energy state, ε is the energy of the state, k is a constant equal to 1.38×10^{-23} J K^{-1}, and T is temperature. Imagine that there are three energy states at 0, 100, and 500 J mol^{-1}.
a) Determine the normalization constant for this distribution.
b) What is the probability of occupying the highest energy state at 298 K?
c) What is the average energy at 298 K?
d) Which state makes the largest contribution to the average energy?

a) Since the energies are given in units of J mol^{-1}, dividing by Avogadro's number will convert this energy to a per particle unit. Alternatively, Avogadro's number can be included with Boltzmann's constant resulting in $k \times N_a = R$, where R = 8.314 J mol^{-1} K^{-1}. Using this relationship:

$$P_{500} = e^{-500 \text{ J mol}^{-1}/(8.314 \text{ J mol}^{-1}\text{ K}^{-1})(298\text{ K})} = 0.817$$
$$P_{100} = e^{-100 \text{ J mol}^{-1}/(8.314 \text{ J mol}^{-1}\text{ K}^{-1})(298\text{ K})} = 0.960$$
$$P_{0} = e^{-0 \text{ J mol}^{-1}/(8.314 \text{ J mol}^{-1}\text{ K}^{-1})(298\text{ K})} = 1$$

Using these probabilities, the normalization constant becomes:

$$A = \frac{1}{\sum_{i=1}^{3} P_i} = \frac{1}{P_0 + P_{100} + P_{500}} \cong 0.360$$

b) With normalization, the normalized probabilities are given by the product of the probabilities determined in part (a) of this question and the normalization constant:

$$P_{500,norm} = A \times P_{500} = 0.360 \times 0.817 = 0.294$$
$$P_{100,norm} = A \times P_{100} = 0.360 \times 0.960 = 0.345$$
$$P_{0,norm} = A \times P_{100} = 0.360$$

c) The average energy is given by:

$$\langle E\rangle=\sum_{i=1}^{3}E_iP_{i,norm}=\left(0\text{ J mol}^{-1}\right)\left(0.360\right)+\left(100\text{ J mol}^{-1}\right)\left(0.345\right)+\left(500\text{ J mol}^{-1}\right)\left(0.294\right)=182\text{ J mol}^{-1}$$

d) Inspection of part (c) of this question illustrates that the highest-energy state makes the largest contribution to the average energy.

P30.22) Consider the following probability distribution corresponding to a particle located between point $x = 0$ and $x = a$:

$$P(x)dx=C\sin^2\left[\frac{\pi x}{a}\right]dx$$

a) Determine the normalization constant, *C*.
b) Determine $\langle x\rangle$.
c) Determine $\langle x^2\rangle$.
d) Determine the variance.

a)

$$1=C\int_0^a\sin^2\left(\frac{\pi x}{a}\right)dx=C\left(\frac{a}{2}\right)$$

$$C=\frac{2}{a}$$

b)

$$\langle x\rangle=\int_0^a(x)\left(\frac{2}{a}\sin^2\left(\frac{\pi x}{a}\right)\right)dx=\frac{2}{a}\int_0^a x\sin^2\left(\frac{\pi x}{a}\right)dx$$

$$=\frac{2}{a}\left(\frac{a^2}{4}\right)=\frac{a}{2}$$

c)

$$\langle x^2\rangle=\int_0^a(x^2)\left(\frac{2}{a}\sin^2\left(\frac{\pi x}{a}\right)\right)dx=\frac{2}{a}\int_0^a x^2\sin^2\left(\frac{\pi x}{a}\right)dx$$

$$=\left(\frac{2}{a}\right)\left[\left(\frac{a^3}{6}\right)-\left(\frac{a^3}{4\pi^2}\right)\right]=a^2\left(\frac{1}{3}-\frac{1}{2\pi^2}\right)$$

d)

$$\sigma^2=\langle x^2\rangle-\langle x\rangle^2=a^2\left(\frac{1}{12}-\frac{1}{2\pi^2}\right)$$

Chapter 31: The Boltzmann Distribution

P31.2) a) Realizing that the most probable outcome from a series of N coin tosses is $N/2$ heads and $N/2$ tails, what is the expression for W_{max} corresponding to this outcome?
b) Given your answer for part (a), derive the following relationship between the weight for an outcome other than the most probable and W_{max}:

$$\log\left(\frac{W}{W_{\max}}\right) = -H\log\left(\frac{H}{N/2}\right) - T\log\left(\frac{T}{N/2}\right)$$

c) We can define the deviation of a given outcome from the most probable outcome using a "deviation index," $\alpha = \frac{H-T}{N}$. Show that the number of heads or tails can be expressed as $H = \frac{N}{2}(1+\alpha)$ and $T = \frac{N}{2}(1-\alpha)$.

d) Finally, demonstrate that $\frac{W}{W_{max}} = e^{-N\alpha^2}$.

a)

$$W = \frac{N!}{H!T!} = \frac{N!}{\left(N/2\right)\left(N/2\right)} = \frac{N!}{\left[\left(N/2\right)\right]^2}$$

b)

$$\ln\left(\frac{W}{W_{\max}}\right) = \ln W - \ln W_{\max} = \ln\left(\frac{N!}{H!T!}\right) - \ln\left(\frac{N!}{\left[\left(N/2\right)!\right]^2}\right)$$

$$= \ln(N!) - \ln(H!) - \ln(T!) - \ln(N!) + 2\ln\left(\left(N/2\right)!\right)$$

$$= -\ln(H!) - \ln(T!) + 2\ln\left(\left(N/2\right)!\right)$$

$$= -H\ln H + H - T\ln T + T + N\ln\left(N/2\right) - N$$

$$= -H\ln H - T\ln T + N\ln\left(N/2\right)$$

$$= -H\ln H - T\ln T + (H+T)\ln\left(N/2\right)$$

$$= -H\ln\left(\frac{H}{N/2}\right) - T\ln\left(\frac{T}{N/2}\right)$$

c) Substituting the definition of part (a) into the expressions for H and T:

$$H = \frac{N}{2}(1+\alpha) = \frac{N}{2}\left(1+\frac{H-T}{N}\right) = \frac{N}{2}+\frac{H-T}{2} = \frac{H+T}{2}+\frac{H-T}{2} = H$$

$$T = \frac{N}{2}(1-\alpha) = \frac{N}{2}\left(1-\frac{H-T}{N}\right) = \frac{N}{2}-\frac{H-T}{2} = \frac{H+T}{2}-\frac{H-T}{2} = T$$

d) Substituting in the result of part (c) into the final equation of part (b):

$$\ln\left(\frac{W}{W_{nax}}\right) = -\frac{N}{2}(1+\alpha)\ln(1+\alpha) - \frac{N}{2}(1-\alpha)\ln(1-\alpha)$$

If $|\alpha| << 1$, then $\ln(1 \pm \alpha) = \pm\alpha$, therefore:

$$\ln\left(\frac{W}{W_{nax}}\right) = -\frac{N}{2}(1+\alpha)\ln(1+\alpha) - \frac{N}{2}(1-\alpha)\ln(1-\alpha)$$

$$= -\frac{N}{2}(1+\alpha)(\alpha) - \frac{N}{2}(1-\alpha)(-\alpha) = -N\alpha^2$$

$$\frac{W}{W_{nax}} = e^{-N\alpha^2}$$

***P31.4*)** Determine the weight associated with the following card hands:

a) Having any five cards
b) Having five cards of the same suit (known as a "flush")

a) The problem can be solved by recognizing that there are 52 total cards ($N = 52$), with 5 cards in the hand ($a_1 = 5$), and 47 out of the hand ($a_0 = 47$):

$$W = \frac{N!}{a_1!a_0!} = \frac{52!}{5!47!} \cong 2.60 \times 10^6$$

b) For an individual suit, there are 13 total cards ($N = 13$), 5 of which must be in the hand ($a_1 = 5$) while the other 8 remain in the deck ($a_0 = 47$). Finally, there are four total suits:

$$W = 4\left(\frac{N!}{a_1!a_0!}\right) = 4\left(\frac{13!}{5!8!}\right) \doteq 5148$$

***P31.7*)** Barometric pressure can be understood using the Boltzmann distribution. The potential energy associated with being a given height above the Earth's surface is *mgh*, where *m* is the mass of the particle of interest, *g* is the acceleration due to gravity, and *h* is height. Using this definition of the potential energy, derive the following expression for pressure:

$$P = P_o e^{-mgh/kT}$$

Assuming that the temperature remains at 298 K, what would you expect the relative pressures of N_2 and O_2 to be at the tropopause, the boundary between the troposphere and stratosphere roughly 11 km above the Earth's surface? At the Earth's surface, the composition of air is roughly 78% N_2, 21% O_2, and the remaining 1% is other gases.

At the Earth's surface, $h = 0$ meters and the total pressure is 1 atm. Using the mole fractions of N_2 and O_2, the partial pressures at the Earth's surface are 0.78 and 0.21 atm, respectively. Given this information, the pressure of N_2 at 11 km is given by:

$$P_{11\text{ km}} = P_{0\text{ km}} e^{-mgh/kT} = (0.78\text{ atm})\, e^{-(0.028\text{ kg mol}^{-1}\times N_A^{-1})(9.8\text{ m s}^{-1})(1.1\times10^{4}\text{ m}) / (1.38\times10^{-23}\text{ J K}^{-1})(298\text{ K})} = 0.230$$

Performing the identical calculation for O_2 yields:

$$P_{11\text{ km}} = P_{0\text{ km}} e^{-mgh/kT} = (0.21\text{ atm})\, e^{-(0.032\text{ kg mol}^{-1}\times N_A^{-1})(9.8\text{ m s}^{-1})(1.1\times10^{4}\text{ m}) / (1.38\times10^{-23}\text{ J K}^{-1})(298\text{ K})} = 0.052$$

P31.9) Consider the following energy-level diagrams, modified from Problem P31.8 by the addition of another excited state with energy of 600 cm^{-1}:
a) At what temperature will the probability of occupying the second energy level be 0.15 for the states depicted in part (a) of the figure?
b) Perform the corresponding calculation for the states depicted in part (b) of the figure.

a)

$$0.15 = p_1 = \frac{e^{-\beta\varepsilon_1}}{q} = \frac{e^{-\beta(300\text{ cm}^{-1})}}{1+e^{-\beta(300\text{ cm}^{-1})}+e^{-\beta(600\text{ cm}^{-1})}}$$

$$0.15+0.15\left(e^{-\beta(300\text{ cm}^{-1})}\right)+0.15\left(e^{-\beta(600\text{ cm}^{-1})}\right) = e^{-\beta(300\text{ cm}^{-1})}$$

$$0.15-0.85\left(e^{-\beta(300\text{ cm}^{-1})}\right)+0.15\left(e^{-\beta(600\text{ cm}^{-1})}\right) = 0$$

The last expression is a quadratic equation with $x = \exp(-\beta(300\text{ cm}^{-1})$. This equation has two roots equal to 0.183 and 5.48. Only the 0.183 root will provide temperature greater than zero, therefore:

$$0.183 = e^{-\beta(300\text{ cm}^{-1})}$$

$$1.70 = \frac{300\text{ cm}^{-1}}{(0.695\text{ cm}^{-1}\text{ K}^{-1})(T)}$$

$$T = 254\text{ K}$$

b)

$$0.15 = p_1 = \frac{2e^{-\beta\varepsilon_1}}{q} = \frac{2e^{-\beta(300\text{ cm}^{-1})}}{1+2e^{-\beta(300\text{ cm}^{-1})}+e^{-\beta(600\text{ cm}^{-1})}}$$

$$0.15+0.30\left(e^{-\beta(300\text{ cm}^{-1})}\right)+0.15\left(e^{-\beta(600\text{ cm}^{-1})}\right) = 2e^{-\beta(300\text{ cm}^{-1})}$$

$$0.15-1.70\left(e^{-\beta(300\text{ cm}^{-1})}\right)+0.15\left(e^{-\beta(600\text{ cm}^{-1})}\right) = 0$$

The last expression is a quadratic equation with $x = \exp(-\beta(300\ \text{cm}^{-1})$. This equation has two roots equal to 0.090 and 11.24. Only the 0.090 root will provide temperature greater than zero, therefore:

$$0.090 = e^{-\beta\left(300\ \text{cm}^{-1}\right)}$$

$$2.41 = \frac{300\ \text{cm}^{-1}}{\left(0.695\ \text{cm}^{-1}\ \text{K}^{-1}\right)(T)}$$

$$T = 179\ \text{K}$$

P31.11) A set of 13 particles occupies states with energies of 0, 100, and 200 cm^{-1}. Calculate the total energy and number of microstates for the following configurations of energy:

a) $a_0 = 8$, $a_1 = 5$, and $a_2 = 0$
b) $a_0 = 9$, $a_1 = 3$, and $a_2 = 1$
c) $a_0 = 10$, $a_1 = 1$, and $a_2 = 2$

Do any of these configurations correspond to the Boltzmann distribution?

The total energy is equal to the sum of energy associated with a given level times the number of particles in that level. For the occupation numbers in a:

$$E = \sum_n \varepsilon_n a_n = \varepsilon_0 a_0 + \varepsilon_1 a_1 + \varepsilon_2 a_2$$

$$= \left(0\ \text{cm}^{-1}\right)(8) + \left(100\ \text{cm}^{-1}\right)(5) + \left(200\ \text{cm}^{-1}\right)(0) = 500\ \text{cm}^{-1}$$

Repeating the calculation for the occupation numbers in (b) and (c) yields the same energy of 500 cm^{-1}. The number of microstates associated with each distribution is given by the weight:

$$W_a = \frac{N!}{\prod_n a_n!} = \frac{N!}{a_0!a_1!a_2!} = \frac{13!}{(8!)(5!)(0!)} = 1287$$

$$W_b = \frac{13!}{(9!)(3!)(1!)} = 2860$$

$$W_c = \frac{13!}{(10!)(1!)(2!)} = 858$$

The ratio of any two occupation numbers for a set of non-degenerate energy levels is given by:

$$\frac{a_i}{a_j} = e^{-\beta\left(\varepsilon_i - \varepsilon_j\right)} = e^{-\left(\frac{\varepsilon_i - \varepsilon_j}{k}\right)\frac{1}{T}}$$

The above expression suggests that the ratio of occupation numbers can be used to determine the temperature. For set (b), comparing the occupation numbers for level 2 and level 0 results in:

$$\frac{a_2}{a_0} = e^{-\left(\frac{\varepsilon_2-\varepsilon_0}{k}\right)\frac{1}{T}}$$

$$\frac{1}{9} = e^{-\left(\frac{200\text{ cm}^{-1}-0\text{ cm}^{-1}}{0.695\text{ cm}^{-1}\text{ K}^{-1}}\right)\frac{1}{T}}$$

$$T = 131\text{ K}$$

Repeating the same calculation for level 1 and level 0:

$$\frac{a_1}{a_0} = e^{-\left(\frac{\varepsilon_2-\varepsilon_0}{k}\right)\frac{1}{T}}$$

$$\frac{3}{9} = e^{-\left(\frac{100\text{ cm}^{-1}-0\text{ cm}^{-1}}{0.695\text{ cm}^{-1}\text{ K}^{-1}}\right)\frac{1}{T}}$$

$$T = 131\text{ K}$$

The distribution of energy in (b) is in accord with the Boltzmann distribution.

P31.14) The ^{13}C nucleus is a spin 1/2 particle as is a proton. However, the energy splitting for a given field strength is roughly 1/4 of that for a proton. Using a 1.45-T magnet as in Example Problem 31.6, what is the ratio of populations in the excited and ground spin states for ^{13}C at 298 K?

Using the information provided in the example problem, the separation in energy is given by:

$$\Delta E = \frac{1}{4}\left(2.82\times10^{-26}\text{ J T}^{-1}\right)B = \frac{1}{4}\left(2.82\times10^{-26}\text{ J T}^{-1}\right)\left(1.45\text{ T}\right) = 1.02\times10^{-26}\text{ J}$$

Using this separation in energy, the ratio in spin-state occupation numbers is:

$$\frac{a_+}{a_-} = e^{-\left(\frac{\varepsilon_+-\varepsilon_-}{kT}\right)} = e^{-\left(\frac{\Delta E}{kT}\right)}$$

$$\frac{a_+}{a_-} = e^{-\left(\frac{1.02\times10^{-26}\text{ J}}{(1.38\times10^{-23}\text{ J K}^{-1})(298\text{ K})}\right)}$$

$$\frac{a_+}{a_-} = 0.999998$$

P31.16) The vibrational frequency of I_2 is 208 cm^{-1}. At what temperature will the population in the first excited state be half that of the ground state?

$$\frac{a_1}{a_0} = \frac{1}{2} = e^{-\beta(\varepsilon_1 - \varepsilon_0)} = e^{-\beta(208\ \text{cm}^{-1})}$$

$$0.5 = e^{\left(\frac{-208\ \text{cm}^{-1}}{(0.695\ \text{cm}^{-1}\ \text{K})(T)}\right)}$$

$$0.693 = \frac{208\ \text{cm}^{-1}}{(0.695\ \text{cm}^{-1}\ \text{K})(T)}$$

$$T = 432\ \text{K}$$

P31.20) The lowest two electronic energy levels of the molecule NO are illustrated here. Determine the probability of occupying one of the higher energy states at 100, 500, and 2000 K.

Both the lower and higher energy states are two-fold degenerate, with an energy spacing of 121.1 cm^{-1}. At 100 K, the partition function is:

$$q = \sum_n e^{-\beta\varepsilon_n} = 2 + e^{-\beta(121.1\ \text{cm}^{-1})} = 2 + e^{-121.1\ \text{cm}^{-1}/(0.695\ \text{cm}^{-1}\ \text{K}^{-1})(100\ \text{K})} = 2.35$$

With the partition function evaluated, the probability of occupying the excited energy level is readily determined:

$$p_1 = \frac{g_1 e^{-\beta\varepsilon_1}}{q} = \frac{2e^{-121.1\ \text{cm}^{-1}/(0.695\ \text{cm}^{-1}\ \text{K}^{-1})(100\ \text{K})}}{2.35} = 0.149$$

At 500 K:

$$q = 2 + e^{-121.1\ \text{cm}^{-1}/(0.695\ \text{cm}^{-1}\ \text{K}^{-1})(500\ \text{K})} = 3.41$$

$$p_1 = \frac{g_1 e^{-\beta\varepsilon_1}}{q} = \frac{2e^{-121.1\ \text{cm}^{-1}/(0.695\ \text{cm}^{-1}\ \text{K}^{-1})(500\ \text{K})}}{3.41} = 0.414$$

Finally, at 2000 K:

$$q = 2 + e^{-121.1\ \text{cm}^{-1}/(0.695\ \text{cm}^{-1}\ \text{K}^{-1})(2000\ \text{K})} = 3.83$$

$$p_1 = \frac{g_1 e^{-\beta\varepsilon_1}}{q} = \frac{2e^{-121.1\ \text{cm}^{-1}/(0.695\ \text{cm}^{-1}\ \text{K}^{-1})(2000\ \text{K})}}{3.83} = 0.479$$

Since there are two states per energy level, the probability of occupying an individual excited state is 1/2 of the above probabilities.

Chapter 32: Ensemble and Molecular Partition Functions

P32.2) Evaluate the translational partition function for Ar confined to a volume of 1000 cm^3 at 298 K. At what temperature will the translational partition function of Ne be identical to that of Ar at 298 K confined to the same volume?

$$q_T(Ar) = \frac{V}{\Lambda^3}$$

$$\Lambda = \left(\frac{h^2}{2\pi mkT}\right)^{1/2} = \left(\frac{(6.626\times10^{-34}\text{ J s})^2}{2\pi\left(\frac{0.0399\text{ kg mol}^{-1}}{\text{N}_\text{A}}\right)(1.38\times10^{-23}\text{ J K}^{-1})(298\text{ K})}\right)^{1/2} = 1.60\times10^{-11}\text{ m}$$

$$q_T(Ar) = \frac{V}{(1.60\times10^{-11}\text{ m})^3} = \frac{(1000\text{ cm}^3)(10^{-6}\text{ m}^3\text{ cm}^{-3})}{(1.60\times10^{-11}\text{ m})^3} = 2.44\times10^{29}$$

If the gases are confined to the same volume, then the partition functions will be equal when the thermal wavelengths are equal:

$$\Lambda(Ne) = \Lambda(Ar) = 1.60\times10^{-11}\text{ m}$$

$$\left(\frac{h^2}{2\pi mkT}\right)^{1/2} = 1.60\times10^{-11}\text{ m}$$

$$T = \frac{h^2}{2\pi mk(1.60\times10^{-11}\text{ m})^2} = \frac{(6.626\times10^{-34}\text{ J s})^2}{2\pi\left(\frac{0.0202\text{ kg mol}^{-1}}{\text{N}_\text{A}}\right)(1.38\times10^{-23}\text{ J K}^{-1})(1.60\times10^{-11}\text{ m})^2}$$

$$T = 590\text{ K}$$

P32.5) For N_2 at 77.3 K, 1 atm, in a 1-cm^3 container, calculate the translational partition function and ratio of this partition function to the number of N_2 molecules present under these conditions.

$$q_T(N_2) = \frac{V}{\Lambda^3}$$

$$\Lambda = \left(\frac{h^2}{2\pi mkT}\right)^{1/2} = \left(\frac{(6.626\times10^{-34}\ \text{J s})^2}{2\pi\left(\frac{0.028\ \text{kg mol}^{-1}}{\text{N}_\text{A}}\right)(1.38\times10^{-23}\ \text{J K}^{-1})(77.3\ \text{K})}\right)^{1/2} = 3.75\times10^{-11}\ \text{m}$$

$$q_T(N_2) = \frac{V}{(3.75\times10^{-11}\ \text{m})^3} = \frac{(1\ \text{cm}^3)(10^{-6}\ \text{m}^3\ \text{cm}^{-3})}{(3.75\times10^{-11}\ \text{m})^3} = 1.90\times10^{25}$$

Next, the number of molecules (N) present at this temperature is determined using the ideal gas law:

$$n = \frac{PV}{RT} = \frac{(1\ \text{atm})(1.00\times10^{-3}\ \text{L})}{(0.0821\ \text{L atm mol}^{-1}\ \text{K}^{-1})(77.3\ \text{K})} = 1.58\times10^{-4}\ \text{mol}$$

$$N = \text{n}\times\text{N}_\text{A} = 9.49\times10^{19}\ \text{molecules}$$

With N, the ratio is readily determined:

$$\frac{q_T}{N} = \frac{1.90\times10^{25}}{9.49\times10^{19}} = 2.00\times10^5$$

P32.8) Consider *para*-H_2 ($B = 60.853$ cm^{-1}) for which only even-J levels are available. Evaluate the rotational partition function for this species at 50 K. Perform this same calculation for HD ($B = 45.655$ cm^{-1}).

For para-H_2, only even J levels are allowed; therefore, the rotational partition function is:

$$q_R = \sum_{J=0,2,4,6,\ldots} (2J+1)e^{-\beta hcBJ(J+1)} = 1 + 5e^{-\frac{(6.626\times10^{-34}\ \text{J s})(3.00\times10^{10}\ \text{m s}^{-1})(60.853\ \text{cm}^{-1})(6)}{(1.38\times10^{-23}\ \text{J K}^{-1})(50\ \text{K})}}$$

$$+ 9e^{-\frac{(6.626\times10^{-34}\ \text{J s})(3.00\times10^{10}\ \text{m s}^{-1})(60.853\ \text{cm}^{-1})(20)}{(1.38\times10^{-23}\ \text{J K}^{-1})(50\ \text{K})}} + \ldots$$

$$= 1 + 1.35\times10^{-4} + \ldots$$

$$\cong 1.00$$

Performing this same calculation for HD where both even and odd J states are allowed:

$$q_R = \sum_{J=0,1,2,3,...} (2J+1)e^{-\beta hcBJ(J+1)} = 1+3e^{-\frac{(6.626\times10^{-34}\text{ J s})(3.00\times10^{10}\text{ m s}^{-1})(45.655\text{ cm}^{-1})(2)}{(1.38\times10^{-23}\text{ J K}^{-1})(50\text{ K})}} + 5e^{-\frac{(6.626\times10^{-34}\text{ J s})(3.00\times10^{10}\text{ m s}^{-1})(45.655\text{ cm}^{-1})(6)}{(1.38\times10^{-23}\text{ J K}^{-1})(50\text{ K})}} + ...$$

$$= 1+0.216+3.74\times10^{-4}+...$$
$$\cong 1.22$$

P32.10) Calculate the rotational partition function for SO_2 at 298 K where $B_A = 2.03$ cm^{-1}, $B_B = 0.344$ cm^{-1}, and $B_C = 0.293$ cm^{-1}.

$$q_R = \frac{\sqrt{\pi}}{\sigma}\left(\frac{1}{\beta B_A}\right)^{1/2}\left(\frac{1}{\beta B_B}\right)^{1/2}\left(\frac{1}{\beta B_C}\right)^{1/2}$$

$$= \frac{\sqrt{\pi}}{2}\left(\frac{(0.695\text{ cm}^{-1})(298\text{ K})}{2.03\text{ cm}^{-1}}\right)^{1/2}\left(\frac{(0.695\text{ cm}^{-1})(298\text{ K})}{0.344\text{ cm}^{-1}}\right)^{1/2}\left(\frac{(0.695\text{ cm}^{-1})(298\text{ K})}{0.293\text{ cm}^{-1}}\right)^{1/2}$$

$$\cong 5832$$

P32.14)

a) Calculate the percent population of the first 10 rotational energy levels for HBr ($B = 8.46$ cm^{-1}) at 298 K.
b) Repeat this calculation for HF assuming that the bond length of this molecule is identical to that of HBr.

Since $T >> \Theta_R$, the high-temperature limit is valid. In this limit, the probability of occupying a specific rotational state (p_J) is:

$$p_J = \frac{(2J+1)e^{-\beta hcBJ(J+1)}}{q} = \frac{(2J+1)e^{-\beta hcBJ(J+1)}}{\left(\frac{1}{\sigma\beta hcB}\right)}$$

Evaluating the above expression for $J = 0$:

$$p_J = \frac{(2J+1)e^{-\beta hcBJ(J+1)}}{\left(\frac{1}{\sigma\beta hcB}\right)} = \frac{1}{\left(\frac{1}{\sigma\beta hcB}\right)}$$

$$= \frac{\sigma hcB}{kT} = \frac{(1)(6.626\times10^{-34}\text{ J s})(3.00\times10^{10}\text{ cm s}^{-1})(8.46\text{ cm}^{-1})}{(1.38\times10^{-23}\text{ J K}^{-1})(298\text{ K})} = 0.041$$

Performing similar calculations for $J = 1$ to 9:

J	p_J	**J**	p_J
0	0.041	5	0.132
1	0.113	6	0.095
2	0.160	7	0.062
3	0.175	8	0.037
4	0.167	9	0.019

a) The rotational constant of HF must be determined before the corresponding level probabilities can be evaluated. The ratio of rotational constants for HBr versus HF yields:

$$\frac{B_{HBr}}{B_{HF}} = \frac{\left(\dfrac{h}{8\pi^2 cI_{HBr}}\right)}{\left(\dfrac{h}{8\pi^2 cI_{HF}}\right)} = \frac{I_{HF}}{I_{HBr}} = \frac{\mu_{HF}r^2}{\mu_{HBr}r^2} = \frac{\mu_{HF}}{\mu_{HBr}} = \frac{\dfrac{m_H m_F}{m_H + m_F}}{\dfrac{m_H m_{Br}}{m_H + m_{Br}}} = 0.962$$

$$B_{HF} = \frac{B_{HBr}}{0.962} = 8.80 \text{ cm}^{-1}$$

With this rotational constant, the p_J values for $J = 0$ to 9 are:

J	p_J	**J**	p_J
0	0.043	5	0.131
1	0.117	6	0.093
2	0.165	7	0.059
3	0.179	8	0.034
4	0.163	9	0.018

b) The rotational constant of HF must be determined before the corresponding level probabilities can be evaluated. The ratio of rotational constants for HBr versus HF yields:

$$\frac{B_{HBr}}{B_{HF}} = \frac{\left(\dfrac{h}{8\pi^2 cI_{HBr}}\right)}{\left(\dfrac{h}{8\pi^2 cI_{HF}}\right)} = \frac{I_{HF}}{I_{HBr}} = \frac{\mu_{HF}r^2}{\mu_{HBr}r^2} = \frac{\mu_{HF}}{\mu_{HBr}} = \frac{\dfrac{m_H m_F}{m_H + m_F}}{\dfrac{m_H m_{Br}}{m_H + m_{Br}}} = 0.962$$

$$B_{HF} = \frac{B_{HBr}}{0.962} = 8.80 \text{ cm}^{-1}$$

With this rotational constant, the p_J values for $J = 0$ to 9 are:

J	p_J	**J**	p_J
0	0.043	5	0.131
1	0.117	6	0.093
2	0.165	7	0.059
3	0.179	8	0.034
4	0.163	9	0.018

P32.18) Evaluate the vibrational partition function for H_2O at 2000 K where the vibrational frequencies are 1615, 3694, and 3802 cm^{-1}.

The total vibrational partition function is the product of partition functions for each vibrational degree of freedom:

$$q_{V,total} = (q_{V,1})(q_{V,2})(q_{V,3})$$

$$= \left(\frac{1}{1-e^{\beta hc\tilde{\nu}_1}}\right)\left(\frac{1}{1-e^{\beta hc\tilde{\nu}_2}}\right)\left(\frac{1}{1-e^{\beta hc\tilde{\nu}_3}}\right)$$

$$= \left(\frac{1}{1-e^{\frac{(6.626\times10^{-34}\text{ J s})(3.00\times10^{10}\text{ cm s}^{-1})(1615\text{ cm}^{-1})}{(1.38\times10^{-23}\text{ J K}^{-1})(2000\text{ K})}}}\right)\left(\frac{1}{1-e^{\frac{(6.626\times10^{-34}\text{ J s})(3.00\times10^{10}\text{ cm s}^{-1})(3694\text{ cm}^{-1})}{(1.38\times10^{-23}\text{ J K}^{-1})(2000\text{ K})}}}\right)$$

$$\times\left(\frac{1}{1-e^{\frac{(6.626\times10^{-34}\text{ J s})(3.00\times10^{10}\text{ cm s}^{-1})(3802\text{ cm}^{-1})}{(1.38\times10^{-23}\text{ J K}^{-1})(2000\text{ K})}}}\right)$$

$$= 1.67$$

P32.20) Evaluate the vibrational partition function for NH_3 at 1000 K for which the vibrational frequencies are 950, 1627.5 (doubly degenerate), 3335, and 3414 cm^{-1} (doubly degenerate). Are there any modes that you can disregard in this calculation? Why or why not?

The total vibrational partition function is the product of partition functions for each vibrational degree of freedom, with the partition function for the mode with degeneracy raised to the power equal to the degeneracy:

$$q_{V,total} = (q_{V,1})(q_{V,2})^2 (q_{V,3})(q_{V,4})^2$$
$$= \left(\frac{1}{1-e^{\beta hc\tilde{\nu}_1}}\right)\left(\frac{1}{1-e^{\beta hc\tilde{\nu}_2}}\right)^2\left(\frac{1}{1-e^{\beta hc\tilde{\nu}_3}}\right)\left(\frac{1}{1-e^{\beta hc\tilde{\nu}_4}}\right)^2$$
$$= \left(\frac{1}{1-e^{\frac{(6.626\times10^{-34}\text{ J s})(3.00\times10^{10}\text{ cm s}^{-1})(950\text{ cm}^{-1})}{(1.38\times10^{-23}\text{ J K}^{-1})(1000\text{ K})}}}\right)\left(\frac{1}{1-e^{\frac{(6.626\times10^{-34}\text{ J s})(3.00\times10^{10}\text{ cm s}^{-1})(1627.5\text{ cm}^{-1})}{(1.38\times10^{-23}\text{ J K}^{-1})(1000\text{ K})}}}\right)^2$$
$$\times\left(\frac{1}{1-e^{\frac{(6.626\times10^{-34}\text{ J s})(3.00\times10^{10}\text{ cm s}^{-1})(3335\text{ cm}^{-1})}{(1.38\times10^{-23}\text{ J K}^{-1})(1000\text{ K})}}}\right)\left(\frac{1}{1-e^{\frac{(6.626\times10^{-34}\text{ J s})(3.00\times10^{10}\text{ cm s}^{-1})(3414\text{ cm}^{-1})}{(1.38\times10^{-23}\text{ J K}^{-1})(1000\text{ K})}}}\right)^2$$
$$= (1.34)(1.11)^2 (1.01)(1.01)^2$$
$$= 1.70$$

Notice that the two highest-frequency vibrational degrees of freedom have partition functions that are near unity; therefore, their contribution to the total partition function is modest, and could be ignored to a reasonable approximation in evaluating the total partition function.

P32.24) Consider a particle free to translate in one dimension. The classical Hamiltonian is $H = \dfrac{p^2}{2m}$.

a) Determine $q_{classical}$ for this system. To what quantum system should you compare it in order to determine the equivalence of the classical and quantum statistical mechanical treatments?
b) Derive $q_{classical}$ for a system with translational motion in three dimensions for which:

$$H = \left(p_x^2 + p_y^2 + p_z^2\right)/2m.$$

a) The particle in a one-dimensional box model is the appropriate quantum-mechanical model for comparison. Integrating

$$q_{class} = \frac{1}{h}\int_{-\infty}^{\infty}\int_{0}^{L} e^{\frac{-\beta p^2}{2m}}\, dx\, dp = \frac{2L}{h}\int_{0}^{\infty} e^{\frac{-\beta p^2}{2m}}\, dp = \frac{2L}{h}\left(\frac{1}{2}\sqrt{\frac{\pi}{\beta/2m}}\right) = L\left(\frac{h^2}{2\pi mkT}\right)^{-1/2} = \frac{L}{\Lambda}$$

b)

$$q_{class} = \frac{1}{h^3}\int_{-\infty}^{\infty}\int_{-\infty}^{\infty}\int_{-\infty}^{\infty}\int_{0}^{L_z}\int_{0}^{L_y}\int_{0}^{L_x} e^{\frac{-\beta\left(p_x^2+p_y^2+p_z^2\right)}{2m}} dxdydzdp_x dp_y dp_z$$

$$= \frac{8\left(L_x L_y L_z\right)}{h^3}\int_0^{\infty}\int_0^{\infty}\int_0^{\infty} e^{\frac{-\beta\left(p_x^2+p_y^2+p_z^2\right)}{2m}} dp_x dp_y dp_z$$

$$= \frac{8V}{h}\left(\frac{1}{2}\sqrt{\frac{\pi}{\beta/2m}}\right)^3 = V\left(\frac{h^2}{2\pi mkT}\right)^{-3/2} = \frac{L}{\Lambda^3}$$

P32.26)

a) Evaluate the electronic partition function for atomic Si at 298 K given the following energy levels:

Level (n)	Energy (cm^{-1})	Degeneracy
0	0	1
1	77.1	3
2	223.2	5
3	6298	5

b) At what temperature will the $n = 3$ energy level contribute 0.1 to the electronic partition function?

a)

$$q_E = \sum_n g_n e^{-\beta\varepsilon_n} = 1e^{-0} + 3e^{-\beta\left(77.1\text{ cm}^{-1}\right)} + 5e^{-\beta\left(223.2\text{ cm}^{-1}\right)} + 5e^{-\beta\left(6298\text{ cm}^{-1}\right)}$$

$$= 1 + 3e^{-\frac{77.1\text{ cm}^{-1}}{\left(0.695\text{ cm}^{-1}\text{ K}^{-1}\right)(298\text{ K})}} + 5e^{-\frac{223.2\text{ cm}^{-1}}{\left(0.695\text{ cm}^{-1}\text{ K}^{-1}\right)(298\text{ K})}} + 3e^{-\frac{6298\text{ cm}^{-1}}{\left(0.695\text{ cm}^{-1}\text{ K}^{-1}\right)(298\text{ K})}}$$

$$= 1 + 3(0.689) + 5(0.340) + 5\left(6.22\times10^{-14}\right)$$

$$q_E = 4.77$$

b) Focusing on the contribution to q_E from the $n = 3$ level:

$$0.1 = g_3 e^{-\beta\varepsilon_3} = 5e^{-\frac{6298\text{ cm}^{-1}}{\left(0.695\text{ cm}^{-1}\text{ K}^{-1}\right)(T)}}$$

$$3.91 = \frac{6298\text{ cm}^{-1}}{\left(0.695\text{ cm}^{-1}\text{ K}^{-1}\right)(T)}$$

$$T = \frac{6298\text{ cm}^{-1}}{\left(0.695\text{ cm}^{-1}\text{ K}^{-1}\right)(3.91)}$$

$$T = 2318\text{ K}$$

P32.28) Determine the total molecular partition function for I_2, confined to a volume of 1000 cm^3 at 298 K. Other information you will find useful: $B = 0.0374\ cm^{-1}$, $\tilde{\nu} = 208\ cm^{-1}$, and the ground electronic state is nondegenerate.

Since q_{Total} is the product of partition functions for each energetic degree of freedom (translational, rotational, vibrational, and elctronic), it is more straightforward to calculate the partition function for each of these degrees of freedom separately, then take the product of these functions:

$$q_T = \frac{V}{\Lambda^3}$$

$$\Lambda = \left(\frac{h^2}{2\pi mkT}\right)^{1/2} = \left(\frac{\left(6.626\times10^{-34}\ \text{J s}\right)^2}{2\pi\left(\frac{0.254\ \text{kg mol}^{-1}}{\text{N}_\text{A}}\right)\left(1.38\times10^{-23}\ \text{J K}^{-1}\right)(298\ \text{K})}\right)^{1/2} = 6.35\times10^{-12}\ \text{m}$$

$$q_T = \frac{V}{\left(6.35\times10^{-12}\ \text{m}\right)^3} = \frac{\left(1000\ \text{cm}^3\right)\left(10^{-6}\ \text{m}^3\ \text{cm}^{-3}\right)}{\left(6.35\times10^{-12}\ \text{m}\right)^3} = 3.91\times10^{30}$$

$$q_R = \left(\frac{1}{\sigma\beta hcB}\right) = \left(\frac{\left(1.38\times10^{-23}\ \text{J K}^{-1}\right)(298\ \text{K})}{(2)\left(6.626\times10^{-34}\ \text{J s}\right)\left(3.00\times10^{10}\ \text{cm s}^{-1}\right)\left(0.0374\ \text{cm}^{-1}\right)}\right)$$
$$= 2.77\times10^3$$

$$q_V = \frac{1}{1-e^{-\beta hc\tilde{\nu}}} = \frac{1}{1-e^{-\frac{\left(6.626\times10^{-34}\ \text{J s}\right)\left(3.00\times10^{10}\ \text{cm s}^{-1}\right)\left(208\ \text{cm}^{-1}\right)}{\left(1.38\times10^{-23}\ \text{J K}^{-1}\right)(298\ \text{K})}}}$$
$$= 1.58$$

$$q_E = 1$$

$$q_{total} = q_T q_R q_V q_E = \left(3.91\times10^{30}\right)\left(2.77\times10^3\right)(1.58)(1) = 1.71\times10^{34}$$

Chapter 33: Statistical Thermodynamics

P33.1) Consider two separate ensembles of particles characterized by the energy-level diagram provided in the text. Derive expressions for the internal energy for each ensemble. At 298 K, which ensemble is expected to have the greatest internal energy?

$$U = -\left(\frac{\partial \ln Q}{\partial \beta}\right)_V = -N\left(\frac{\partial \ln q}{\partial \beta}\right)_V = \frac{-N}{q}\left(\frac{\partial q}{\partial \beta}\right)_V$$

$$q_A = \sum_n g_n e^{-\beta\varepsilon_n} = 1 + e^{-\beta(300\ \mathrm{cm^{-1}})} + e^{-\beta(600\ \mathrm{cm^{-1}})}$$

$$\begin{aligned}
U_A &= \frac{-N}{1+e^{-\beta(300\ \mathrm{cm^{-1}})}+e^{-\beta(600\ \mathrm{cm^{-1}})}}\left(\frac{\partial}{\partial\beta}\left(1+e^{-\beta(300\ \mathrm{cm^{-1}})}+e^{-\beta(600\ \mathrm{cm^{-1}})}\right)\right)\\
&= \frac{-N}{1+e^{-\beta(300\ \mathrm{cm^{-1}})}+e^{-\beta(600\ \mathrm{cm^{-1}})}}\left(-\left(\left(300\ \mathrm{cm^{-1}}\right)e^{-\beta(300\ \mathrm{cm^{-1}})}+\left(600\ \mathrm{cm^{-1}}\right)e^{-\beta(600\ \mathrm{cm^{-1}})}\right)\right)\\
&= \frac{N\left(\left(300\ \mathrm{cm^{-1}}\right)e^{-\beta(300\ \mathrm{cm^{-1}})}+\left(600\ \mathrm{cm^{-1}}\right)e^{-\beta(600\ \mathrm{cm^{-1}})}\right)}{1+e^{-\beta(300\ \mathrm{cm^{-1}})}+e^{-\beta(600\ \mathrm{cm^{-1}})}}
\end{aligned}$$

$$q_B = \sum_n g_n e^{-\beta\varepsilon_n} = 1 + 2e^{-\beta(300\ \mathrm{cm^{-1}})} + e^{-\beta(600\ \mathrm{cm^{-1}})}$$

$$\begin{aligned}
U_B &= \frac{-N}{1+2e^{-\beta(300\ \mathrm{cm^{-1}})}+e^{-\beta(600\ \mathrm{cm^{-1}})}}\left(\frac{\partial}{\partial\beta}\left(1+2e^{-\beta(300\ \mathrm{cm^{-1}})}+e^{-\beta(600\ \mathrm{cm^{-1}})}\right)\right)\\
&= \frac{-N}{1+2e^{-\beta(300\ \mathrm{cm^{-1}})}+e^{-\beta(600\ \mathrm{cm^{-1}})}}\left(-\left(\left(600\ \mathrm{cm^{-1}}\right)e^{-\beta(300\ \mathrm{cm^{-1}})}+\left(600\ \mathrm{cm^{-1}}\right)e^{-\beta(600\ \mathrm{cm^{-1}})}\right)\right)\\
&= \frac{N\left(600\ \mathrm{cm^{-1}}\right)\left(e^{-\beta(300\ \mathrm{cm^{-1}})}+e^{-\beta(600\ \mathrm{cm^{-1}})}\right)}{1+2e^{-\beta(300\ \mathrm{cm^{-1}})}+e^{-\beta(600\ \mathrm{cm^{-1}})}}
\end{aligned}$$

Evaluating the expressions for U_A and U_B at 298 K:

$$U_A = \frac{N\left(\left(300 \text{ cm}^{-1}\right) e^{-\frac{\left(300 \text{ cm}^{-1}\right)}{\left(0.695 \text{ cm}^{-1} \text{ K}^{-1}\right)(298 \text{ K})}} + \left(600 \text{ cm}^{-1}\right) e^{-\frac{\left(600 \text{ cm}^{-1}\right)}{\left(0.695 \text{ cm}^{-1} \text{ K}^{-1}\right)(298 \text{ K})}} \right)}{1 + e^{-\frac{\left(300 \text{ cm}^{-1}\right)}{\left(0.695 \text{ cm}^{-1} \text{ K}^{-1}\right)(298 \text{ K})}} + e^{-\frac{\left(600 \text{ cm}^{-1}\right)}{\left(0.695 \text{ cm}^{-1} \text{ K}^{-1}\right)(298 \text{ K})}}}$$
$$= N\left(80.3 \text{ cm}^{-1}\right)$$

$$U_B = \frac{N\left(600 \text{ cm}^{-1}\right)\left(e^{-\frac{\left(300 \text{ cm}^{-1}\right)}{\left(0.695 \text{ cm}^{-1} \text{ K}^{-1}\right)(298 \text{ K})}} + e^{-\frac{\left(600 \text{ cm}^{-1}\right)}{\left(0.695 \text{ cm}^{-1} \text{ K}^{-1}\right)(298 \text{ K})}} \right)}{1 + 2e^{-\frac{\left(300 \text{ cm}^{-1}\right)}{\left(0.695 \text{ cm}^{-1} \text{ K}^{-1}\right)(298 \text{ K})}} + e^{-\frac{\left(600 \text{ cm}^{-1}\right)}{\left(0.695 \text{ cm}^{-1} \text{ K}^{-1}\right)(298 \text{ K})}}}$$
$$= N\left(114 \text{ cm}^{-1}\right)$$

Ensemble B will have the larger internal energy.

P33.6) Consider an ensemble of units in which the first excited electronic state at energy ε_1 is m_1-fold degenerate, and the energy of the ground state is m_o-fold degenerate with energy ε_0.

a) Demonstrate that if $\varepsilon_0 = 0$, the expression for the electronic partition function is

$$q_E = m_0 \left(1 + \frac{m_1}{m_0} e^{-\varepsilon_1 / kT} \right)$$

b) Determine the expression for the internal energy U of an ensemble of N such units. What is the limiting value of U as the temperature approaches zero and infinity?

a)

$$q = m_0 e^{-\beta\varepsilon_0} + m_1 e^{-\beta\varepsilon_1} = m_0 + m_1 e^{-\beta\varepsilon_1}$$
$$= m_0 + m_0 \left(\frac{m_1}{m_0} \right) e^{-\beta\varepsilon_1}$$
$$= m_0 \left(1 + \left(\frac{m_1}{m_0} \right) e^{-\beta\varepsilon_1} \right)$$
$$= m_0 \left(1 + \left(\frac{m_1}{m_0} \right) e^{-\frac{\varepsilon_1}{kT}} \right)$$

b)

$$U = \frac{-N}{q}\left(\frac{\partial q}{\partial \beta}\right)_V = \frac{-N}{q}\left(\frac{\partial}{\partial \beta}\left(m_0\left(1+\left(\frac{m_1}{m_0}\right)e^{-\beta\varepsilon_1}\right)\right)\right)_V$$

$$= \frac{-N}{q}\left(-m_1\varepsilon_1 e^{-\beta\varepsilon_1}\right)$$

$$= \frac{Nm_1\varepsilon_1 e^{-\beta\varepsilon_1}}{m_0\left(1+\left(\frac{m_1}{m_0}\right)e^{-\beta\varepsilon_1}\right)} = \frac{Nm_1\varepsilon_1 e^{-\frac{\varepsilon_1}{kT}}}{m_0\left(1+\left(\frac{m_1}{m_0}\right)e^{-\frac{\varepsilon_1}{kT}}\right)}$$

Looking at the limiting behavior with temperature:

$$\lim_{T\to 0} U == \lim_{T\to 0}\frac{Nm_1\varepsilon_1 e^{-\frac{\varepsilon_1}{kT}}}{m_0\left(1+\left(\frac{m_1}{m_0}\right)e^{-\frac{\varepsilon_1}{kT}}\right)} = \lim_{T\to 0}\frac{Nm_1\varepsilon_1}{m_0\left(e^{\frac{\varepsilon_1}{kT}}+\left(\frac{m_1}{m_0}\right)\right)} = 0$$

$$\lim_{T\to \infty} U == \lim_{T\to 0}\frac{Nm_1\varepsilon_1 e^{-\frac{\varepsilon_1}{kT}}}{m_0\left(1+\left(\frac{m_1}{m_0}\right)e^{-\frac{\varepsilon_1}{kT}}\right)} = \frac{Nm_1\varepsilon_1}{m_0+m_1}$$

P33.9) Determine the vibrational contribution to C_V for a mole of HCl ($\tilde{\nu} = 2886\ \text{cm}^{-1}$) over a temperature range from 500 to 5000 K in 500-K intervals and plot your result. At what temperature do you expect to reach the high-temperature limit for the vibrational contribution to C_V?

The problem requires evaluation of the following expression versus temperature:

$$C_V = \frac{N}{kT^2}\left(hc\tilde{\nu}\right)^2 \frac{e^{\frac{hc\tilde{\nu}}{kT}}}{\left(e^{\frac{hc\tilde{\nu}}{kT}}-1\right)^2}$$

Using Excel or a similar program, the following plot of the molar heat capacity versus temperature can be constructed.

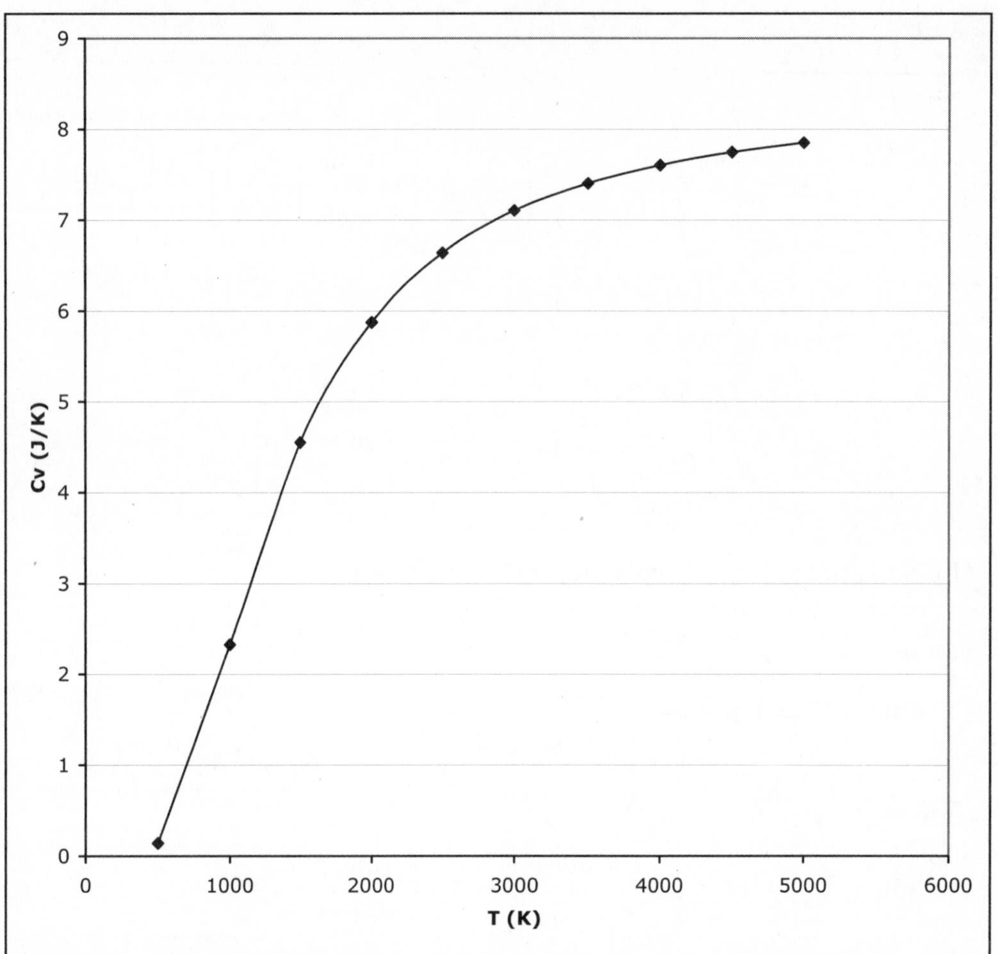

The high-temperature limit value for the molar heat capacity is (1 mol) × R = 8.314 J K^{-1}. Comparison of this value to the graph illustrates that the high-temperature limit will be value for temperatures well above 6000. Specifically, the high-temperature limit is applicable when T > 10Θ_V or ~40,000 K for Hcl.

P33.12) The speed of sound is given by the relationship

$$c_{sound} = \left(\frac{\frac{C_p}{C_V} RT}{M} \right)^{1/2}$$

where C_p is the constant pressure heat capacity (equal to C_V +R), R is the ideal gas constant, T is temperature, and M is molar mass.

a) What is the expression for the speed of sound for an ideal monatomic gas?
b) What is the expression for the speed of sound of an ideal diatomic gas?

c) What is the speed of sound in air at 298 K, assuming that air is mostly made up of nitrogen ($B = 2.00$ cm^{-1} and $\tilde{\nu} = 2359$ cm^{-1})?

a) For a monatomic gas, only translational degrees of freedom contribute to the C_V:

$$C_V = \frac{3}{2}Nk = \frac{3}{2}nR$$

$$\frac{C_P}{C_V} = \frac{\frac{5}{2}nR}{\frac{3}{2}nR} = \frac{5}{3}$$

$$c_{sound} = \left(\frac{\frac{C_p}{C_V}RT}{M}\right)^{1/2} = \left(\frac{\frac{5}{3}RT}{M}\right)^{1/2}$$

b) In addition to translations, rotational and vibrational degrees of freedom will also contribute to C_V:

$$C_V = C_{V,trans} + C_{V,rot} + C_{V,vib} = \frac{3}{2}nR + nR + nR\beta^2 \left(hc\tilde{\nu}\right)^2 \frac{e^{\beta hc\tilde{\nu}}}{\left(e^{\beta hc\tilde{\nu}}+1\right)^2}$$

$$= nR\left(\frac{5}{2} + \beta^2 \left(hc\tilde{\nu}\right)^2 \frac{e^{\beta hc\tilde{\nu}}}{\left(e^{\beta hc\tilde{\nu}}+1\right)^2}\right)$$

$$\frac{C_P}{C_V} = \frac{\left(\frac{7}{2} + \beta^2 \left(hc\tilde{\nu}\right)^2 \frac{e^{\beta hc\tilde{\nu}}}{\left(e^{\beta hc\tilde{\nu}}+1\right)^2}\right)}{\left(\frac{5}{2} + \beta^2 \left(hc\tilde{\nu}\right)^2 \frac{e^{\beta hc\tilde{\nu}}}{\left(e^{\beta hc\tilde{\nu}}+1\right)^2}\right)}$$

$$c_{sound} = \left(\frac{\frac{C_p}{C_V} RT}{M}\right)^{1/2} = \left(\frac{\left(\frac{\left(\frac{7}{2} + \beta^2 \left(hc\tilde{\nu}\right)^2 \frac{e^{\beta hc\tilde{\nu}}}{\left(e^{\beta hc\tilde{\nu}}+1\right)^2}\right)}{\left(\frac{5}{2} + \beta^2 \left(hc\tilde{\nu}\right)^2 \frac{e^{\beta hc\tilde{\nu}}}{\left(e^{\beta hc\tilde{\nu}}+1\right)^2}\right)}\right) RT}{M}\right)^{1/2}$$

c) First, evaluation of the vibrational contribution to C_V demonstrates that this contribution is small relative to the contribution from translational and rotational degrees of freedom, and can be neglected to good approximation:

$$\frac{C_{V,vib}}{nR} = \beta^2 \left(hc\tilde{\nu}\right)^2 \frac{e^{\beta hc\tilde{\nu}}}{\left(e^{\beta hc\tilde{\nu}}+1\right)^2}$$

$$= \left(\frac{\left(6.626\times10^{-34}\ \text{J s}\right)\left(3.00\times10^{10}\ \text{cm s}^{-1}\right)\left(2359\ \text{cm}^{-1}\right)}{\left(1.38\times10^{-23}\ \text{J K}^{-1}\right)\left(298\ \text{K}\right)}\right)^2$$

$$\times \frac{e^{\frac{\left(6.626\times10^{-34}\ \text{J s}\right)\left(3.00\times10^{10}\ \text{cm s}^{-1}\right)\left(2359\ \text{cm}^{-1}\right)}{\left(1.38\times10^{-23}\ \text{J K}^{-1}\right)\left(298\ \text{K}\right)}}}{\left(e^{\frac{\left(6.626\times10^{-34}\ \text{J s}\right)\left(3.00\times10^{10}\ \text{cm s}^{-1}\right)\left(2359\ \text{cm}^{-1}\right)}{\left(1.38\times10^{-23}\ \text{J K}^{-1}\right)\left(298\ \text{K}\right)}}+1\right)^2}$$

$$= 1.47\times10^{-3} << \frac{5}{2}$$

Therefore, keeping only the translational and rotational contributions to C_V yields the following value for the speed of sound in N_2:

$$c_{sound} = \left(\frac{\frac{C_p}{C_V}RT}{M}\right)^{1/2} \cong \left(\frac{\frac{7}{5}RT}{M}\right)^{1/2} = \left(\frac{\frac{7}{5}\left(8.314\ \text{J mol}^{-1}\ \text{K}^{-1}\right)\left(298\ \text{K}\right)}{\left(0.028\ \text{kg mol}^{-1}\right)}\right)^{1/2}$$

$$= 352\ \text{m s}^{-1}$$

P33.15) Determine the molar entropy for 1 mol of gaseous Ar at 200, 300, and 500 K and $V = 1000\ \text{cm}^3$ assuming that Ar can be treated as an ideal gas. How does the result of this calculation change if the gas is Kr instead of Ar?

Determining the molar entropy for gaseous Ar at 200 K yields:

$$S = \frac{U}{T} + k\ln Q = \frac{3}{2}Nk + k\ln\left(\frac{q^N}{N!}\right) = \frac{3}{2}Nk + NK\ln q - k\ln(N!)$$

$$= \frac{3}{2}Nk + Nk\ln q - k(N\ln N - N)$$

$$= \frac{5}{2}nR + nR\ln q - nR\ln(nN_A)$$

$$S_m = \frac{5}{2}R + R\ln\left(\frac{V}{\Lambda^3}\right) - R\ln(6.022\times10^{23})$$

$$= 20.79\ \text{J mol}^{-1}\ \text{K}^{-1} + R\ln\left(\frac{V}{\left(\frac{h^2}{2\pi mkT}\right)^{3/2}}\right) - 456\ \text{J mol}^{-1}\ \text{K}^{-1}$$

$$= 20.79\ \text{J mol}^{-1}\ \text{K}^{-1}$$

$$+ R\ln\left(\frac{0.001\ \text{m}^3}{N_A\left(\frac{(6.626\times10^{-34}\ \text{J s})^2}{2\pi\left(\frac{0.040\ \text{kg mol}^{-1}}{N_A}\right)(1.38\times10^{-23}\ \text{J K}^{-1})(200\ \text{K})}\right)^{3/2}}\right) - 456\ \text{J mol}^{-1}\ \text{K}^{-1}$$

$$= 123\ \text{J mol}^{-1}\ \text{K}^{-1}$$

Repeating the calculation for T = 300 and 500 K, and the molar entropy is found to be 128 J mol^{-1} K^{-1} and 135 J mol^{-1} K^{-1}, respectively. Kr is heavier than Ar; therefore, the thermal wavelength will be shorter, and the translational partition function will correspondingly be larger. Since the molar entropy is linear related to ln(q), we would expected the molar entropy for Kr to be greater than that of Ar. This expectation can be confirmed by repeating the above calculation for Kr, or by simply looking at the difference in entropy between Kr and Ar:

$$S_{Kr} - S_{Ar} = \left(\frac{U}{T} + k\ln Q\right)_{Kr} - \left(\frac{U}{T} + k\ln Q\right)_{Ar}$$

$$= k\ln\left(\frac{Q_{Kr}}{Q_{Ar}}\right) = Nk\ln\left(\frac{q_{Kr}}{q_{Ar}}\right) = Nk\ln\left(\frac{\Lambda_{Ar}^3}{\Lambda_{Kr}^3}\right) = Nk\ln\left(\frac{m_{Kr}}{m_{Ar}}\right)^{3/2}$$

$$S_{m,Kr} - S_{m,Ar} = \frac{3}{2}R\ln\left(\frac{m_{Kr}}{m_{Ar}}\right) = \frac{3}{2}R\ln(2.00)$$

$$S_{m,Kr} = 8.64 \text{ J mol}^{-1}\text{ K}^{-1} + S_{m,Ar}$$

P33.17) Determine the standard molar entropy of N_2O, a linear triatomic molecule at 298 K. For this molecule, $B = 0.419 \text{ cm}^{-1}$ and $\tilde{\nu}_1 = 1285 \text{ cm}^{-1}$, $\tilde{\nu}_2 = 589 \text{ cm}^{-1}$ (doubly degenerate), and $\tilde{\nu}_3 = 2224 \text{ cm}^{-1}$.

At 298.15 K, both the translational and rotational degrees of freedom will be in the high-temperature limit, but the vibrational contributions must be calculated. Using the standard volume of 24.5 L and temperature of 298.15 K the entropy is determined as follows:

$$U_m^\circ = U_{T,m}^\circ + U_{R,m}^\circ + U_{V,m}^\circ + U_{E,m}^\circ$$
$$= \frac{3}{2}RT + RT + N_A hc\left[\left(\frac{\tilde{\nu}_1}{e^{\beta hc\tilde{\nu}_1}-1}\right) + 2\left(\frac{\tilde{\nu}_2}{e^{\beta hc\tilde{\nu}_2}-1}\right) + \left(\frac{\tilde{\nu}_3}{e^{\beta hc\tilde{\nu}_3}-1}\right)\right]$$
$$= \frac{5}{2}R(298.15\text{ K}) + 906\text{ J mol}^{-1} = 7.10\text{ kJ mol}^{-1}$$

$$S_m^\circ = \frac{U_m^\circ}{T} + k\ln Q = \frac{U_m^\circ}{T} + k\ln\left(\frac{q_{total}^N}{N!}\right) = \frac{U_m^\circ}{T} + Nk\ln(q_{total}) - k\ln(N!)$$
$$= \frac{7.10\text{ kJ mol}^{-1}}{298.15\text{ K}} + R\ln(q_T q_R q_V q_E) - R\ln((1\text{ mol})N_A) + R$$
$$= -424\text{ J mol}^{-1}\text{ K}^{-1} + R\ln(q_T q_R q_V q_E)$$

$$q_T = \left(\frac{V}{\Lambda^3}\right) = \frac{0.0245\text{ m}^3}{3.55\times10^{-33}\text{ m}^3} = 6.91\times10^{30}$$
$$q_R = \left(\frac{kT}{\sigma B}\right) = \frac{(0.695\text{ cm}^{-1}\text{ K}^{-1})(298.15\text{ K})}{0.42\text{ cm}^{-1}} = 493$$
$$q_V = \left(\frac{1}{1-e^{-\beta\tilde{\nu}_1}}\right)\left(\frac{1}{1-e^{-\beta\tilde{\nu}_2}}\right)^2\left(\frac{1}{1-e^{-\beta\tilde{\nu}_3}}\right)$$
$$= \left(\frac{1}{1-e^{-\frac{1285\text{ cm}^{-1}}{(0.695\text{ cm}^{-1})(298.15\text{ K})}}}\right)\left(\frac{1}{1-e^{-\frac{589\text{ cm}^{-1}}{(0.695\text{ cm}^{-1})(298.15\text{ K})}}}\right)^2\left(\frac{1}{1-e^{-\frac{2224\text{ cm}^{-1}}{(0.695\text{ cm}^{-1})(298.15\text{ K})}}}\right)$$
$$= 1.00$$
$$q_E = 1.00$$

$$S_m^\circ = -424\text{ J mol}^{-1}\text{ K}^{-1} + R\ln(q_T q_R q_V q_E) = -424\text{ J mol}^{-1}\text{ K}^{-1} + 643\text{ J mol}^{-1}\text{ K}^{-1}$$
$$S_m^\circ = 219\text{ J mol}^{-1}\text{ K}^{-1}$$

P33.22) The molecule NO has a ground electronic level that is doubly degenerate, and a first excited level at 121.1 cm^{-1} that is also twofold degenerate. Determine the contribution of electronic degrees of freedom to the standard molar entropy of NO. Compare your result to $R\ln(4)$. What is the significance of this comparison?

$$q_E = g_0 + g_1 e^{-\beta\varepsilon_1} = 2 + 2e^{-\beta hc(121.1\text{ cm}^{-1})} = 3.11$$

$$U_E = \frac{-N}{q_E}\left(\frac{\partial q_E}{\partial \beta}\right)_V = \frac{2Nhc(121.1\text{ cm}^{-1})e^{-\beta hc(121.1\text{ cm}^{-1})}}{2+2e^{-\beta hc(121.1\text{ cm}^{-1})}}$$

$$= \frac{Nhc(121.1\text{ cm}^{-1})e^{-\beta hc(121.1\text{ cm}^{-1})}}{1+e^{-\beta hc(121.1\text{ cm}^{-1})}}$$

$$U_{E,m} = \frac{N_A hc(121.1\text{ cm}^{-1})e^{-\beta hc(121.1\text{ cm}^{-1})}}{1+e^{-\beta hc(121.1\text{ cm}^{-1})}} = 518\text{ J mol}^{-1}$$

$$S_{E,m} = \frac{U_{E,m}}{T} + R\ln(q_E) = \frac{518\text{ J mol}^{-1}}{298.15\text{ K}} + R\ln(3.11)$$

$$= 1.73\text{ J mol}^{-1}\text{ K}^{-1} + 9.43\text{ J mol}^{-1}\text{ K}^{-1}$$

$$= 11.2\text{ J mol}^{-1}\text{ K}^{-1}$$

$$R\ln(4) = 11.5\text{ J mol}^{-1}\text{ K}^{-1}$$

At sufficiently high temperatures, $q_E = 4$, and the contribution of the two lowest electronic energy levels to the molar entropy would equal $R\ln(4)$. The similarity between the calculated and limiting values demonstrates that this limiting behavior is being approached at this temperature.

P33.27) Calculate the standard Helmholtz energy for molar ensembles of Ne and Kr at 298 K.

First, performing the calculation for Ne (M = 0.020 kJ mol^{-1}):

$$A = -kT\ln Q = -kT\ln\left(\frac{q^N}{N!}\right)$$

$$= -NkT\ln q + kT\ln(N!)$$

$$= -NkT\ln q + kT(N\ln N - N)$$

$$= -NkT(\ln q - \ln N + 1)$$

$$q = q_T = \frac{V}{\Lambda^3} = \frac{0.0245\text{ m}^3}{\left(\frac{h^2}{2\pi mkT}\right)^{3/2}} = \frac{0.0245\text{ m}^3}{1.08\times10^{-32}\text{ m}^3} = 2.27\times10^{30}$$

$$N = n\times N_A = 6.022\times10^{23}$$

$$A = -nRT(\ln q - \ln N + 1)$$

$$A_m^\circ = -(8.314\text{ J mol}^{-1}\text{ K}^{-1})(298.15\text{ K})\left(\ln(2.27\times10^{30}) - \ln(6.022\times10^{23}) + 1\right)$$

$$A_m^\circ = -40.0\text{ kJ mol}^{-1}$$

This calculation can be repeated for Kr (M = 0.083 kJ mol^{-1}), or the difference between the Helmholtz energy of Kr and Ne can be determined:

$$
\begin{aligned}
A_{Kr} - A_{Ar} &= -kT\left(\ln Q_{Kr} - \ln Q_{Ar}\right) \\
&= -kT \ln\left(\frac{Q_{Kr}}{Q_{Ar}}\right) \\
&= -NkT \ln\left(\frac{q_{Kr}}{q_{Ar}}\right) \\
&= -NkT \ln\left(\left(\frac{\Lambda_{Ar}}{\Lambda_{Kr}}\right)^{3}\right) \\
&= -NkT \ln\left(\left(\frac{m_{Kr}}{m_{Ar}}\right)^{3/2}\right) \\
&= -\frac{3}{2}NkT \ln\left(\frac{m_{Kr}}{m_{Ar}}\right) \\
&= -\frac{3}{2}nRT \ln\left(\frac{0.083}{0.020}\right) \\
A^{\circ}_{Kr,m} &= -\frac{3}{2}RT \ln(4.15) + A^{\circ}_{Ar,m} = -40.3 \text{ kJ mol}^{-1}
\end{aligned}
$$

P33.31) Determine the equilibrium constant for the dissociation of sodium at 298 K: $Na_2(g) \rightleftarrows 2Na(g)$.

For Na_2, $B = 0.155 \text{ cm}^{-1}$, $\tilde{\nu} = 159 \text{ cm}^{-1}$, the dissociation energy is 70.4 kJ/mol, and the ground-state electronic degeneracy for Na is 2.

$$K = \frac{\left(\frac{q}{N_A}\right)^2_{Na}}{\left(\frac{q}{N_A}\right)_{Na_2}} e^{-\beta \varepsilon_D}$$

$$q_{Na} = q_T q_E = \left(\frac{V}{\Lambda^3}\right)(2) = \left(\frac{0.0245 \text{ m}^3}{9.38\times10^{-33} \text{ m}^3}\right)(2) = 5.22\times10^{30}$$

$$q_{Na_2} = q_T q_R q_V q_E = \left(\frac{V}{\Lambda^3}\right)\left(\frac{kT}{\sigma B}\right)\left(\frac{1}{1-e^{-\beta hc\tilde{\nu}}}\right)(1)$$

$$= \left(\frac{0.0245 \text{ m}^3}{3.32\times10^{-33} \text{ m}^3}\right)\left(\frac{\left(0.695 \text{ cm}^{-1} \text{ K}^{-1}\right)(298.15 \text{ K})}{(2)\left(0.155 \text{ cm}^{-1}\right)}\right)$$

$$\times\left(\frac{1}{1-e^{-\frac{\left(6.626\times10^{-34} \text{ J s}\right)\left(3.00\times10^{10} \text{ cm s}^{-1}\right)\left(159 \text{ cm}^{-1}\right)}{\left(1.38\times10^{-23} \text{ J K}^{-1}\right)(298.15 \text{ K})}}}\right)(1)$$

$$= 9.22\times10^{33}$$

$$K = \frac{\left(\frac{5.22\times10^{30}}{N_A}\right)^2_{Na}}{\left(\frac{9.22\times10^{33}}{N_A}\right)_{Na_2}} e^{-\frac{70{,}400 \text{ J mol}^{-1}/N_A}{\left(1.38\times10^{-23} \text{ J K}^{-1}\right)(298.15 \text{ K})}}$$

$$= 2.25\times10^{-9}$$

Chapter 34: Kinetic Theory of Gases

P34.1) Consider a collection of gas particles confined to translate in two dimensions (for example, a gas molecule on a surface). Derive the Maxwell speed distribution for such a gas.

Beginning with the Maxwell-Boltzmann velocity distribution in one-dimension

$$f(v_j) = \left(\frac{m}{2\pi kT}\right)^{1/2} e^{-\frac{m}{2kT}v_j^2}$$

and the definition of speed in 2 dimensions

$$\nu = \left(v_x^2 + v_y^2\right)^{1/2}$$

The speed distribution in 2 dimensions is given by:

$$F d\nu = f(v_x) f(v_y) dv_x dv_y$$

where dv_j is the differential of velocity in the jth direction. Thus

$$\begin{aligned} F d\nu &= \left(\frac{m}{2\pi kT}\right)^{1/2} e^{-\frac{m}{2kT}v_x^2} \left(\frac{m}{2\pi kT}\right)^{1/2} e^{-\frac{m}{2kT}v_y^2} dv_x dv_y \\ &= \left(\frac{m}{2\pi kT}\right) e^{-\frac{m}{2kT}\left(v_x^2+v_y^2\right)} dv_x dv_y \\ F d\nu &= \left(\frac{m}{2\pi kT}\right) e^{-\frac{m}{2kT}\nu^2} dv_x dv_y \end{aligned}$$

The differential area is defined as:

$$dv_x dv_y = 2\pi\nu d\nu$$

Substituting this into the expression for $Fd\nu$:

$$F d\nu = \left(\frac{m}{2\pi kT}\right) e^{-\frac{m}{2kT}\nu^2} (2\pi\nu) d\nu = \frac{m}{kT} e^{-\frac{m}{2kT}\nu^2}$$

P34.4) Compare the average speed and average kinetic energy of O_2 with that of CCl_4 at 298 K.

$$\mathrm{M}_{\mathrm{O}_2} = 0.032 \text{ kg mol}^{-1} \qquad \mathrm{M}_{\mathrm{CCl}_4} = 0.154 \text{ kg mol}^{-1}$$

At the same temperature, the speed for two particles of different mass is related by the square root of the mass ratios. For this case of O_2 and CCl_4:

$$v_{\mathrm{CCl}_4} = \left(\frac{\mathrm{M}_{\mathrm{O}_2}}{\mathrm{M}_{\mathrm{CCl}_4}}\right)^{1/2} v_{\mathrm{O}_2}$$

$$v_{\mathrm{CCl}_4} = \left(\frac{0.0320 \text{ kg mol}^{-1}}{0.154 \text{ kg mol}^{-1}}\right)^{1/2} v_{\mathrm{O}_2} = (0.456)v_{\mathrm{O}_2}$$

The average speed for O_2 at 298 K is:

$$v_{\text{ave}} = \sqrt{\frac{8RT}{\pi M}} = \sqrt{\frac{8\left(8.314 \text{ J mol}^{-1}\text{ K}^{-1}\right)(298 \text{ K})}{\pi \cdot 0.032 \text{ kg mol}^{-1}}} = 444 \text{ m s}^{-1}$$

Using this result, the average speed for CCl_4 at this same temperature is:

$$v_{\text{ave,CCl}_4} = 0.456 \times v_{\text{ave, O}_2}$$

$$v_{\text{ave,CCl}_4} = (0.456)\left(444 \text{ m s}^{-1}\right)$$

$$v_{\text{ave,CCl}_4} = 203 \text{ m s}^{-1}$$

The average kinetic energy is mass independent, thus for a given temperature, all gases have the same kinetic energy. The average kinetic energy per gas particle is therefore:

$$\langle \mathrm{KE} \rangle = \frac{3}{2}\mathrm{kT} = \frac{3}{2}\left(1.38\times10^{-23}\text{ J K}^{-1}\right)(298 \text{ K})$$

$$\langle \mathrm{KE} \rangle = 6.17\times10^{-21}\text{ J}$$

***P34.8*)** The probability that a particle will have a velocity in the x direction in the range of $-\mathrm{v}_{x_0}$ and v_{x_0} is given by

$$f(-\mathrm{v}_{x_0} \le \mathrm{v}_x \le \mathrm{v}_{x_0}) = \left(\frac{m}{2\pi kT}\right)^{1/2} \int_{-\mathrm{v}_{x0}}^{\mathrm{v}_{x0}} e^{\frac{-m\mathrm{v}_x^2}{2kT}} d\mathrm{v}_x = \left(\frac{2m}{\pi kT}\right)^{1/2} \int_0^{\mathrm{v}_{x0}} e^{\frac{-m\mathrm{v}_x^2}{2kT}} d\mathrm{v}_x$$

The preceding integral can be rewritten using the following substitution: $\xi^2 = \dfrac{m\mathrm{v}_x^2}{2kT}$,

resulting in $f(-\mathrm{v}_{x_0} \le \mathrm{v}_x \le \mathrm{v}_{x_0}) = \dfrac{2}{\sqrt{\pi}}\displaystyle\int_0^{\xi_0} e^{-\xi^2} d\xi$, which can be evaluated using the error function defined as $\mathrm{erf}(z)=\frac{2}{\sqrt{\pi}}\int_0^z e^{-x^2}dx$. The complementary error function is defined as $\mathrm{erfc}(z)=1-\mathrm{erf}(z)$.. Finally, a plot of both erf(z) and erfc(z) as a function of z is shown in the text (tabulated values are available in the Math Supplement, Appendix B): Using this graph of erf(z), determine the probability that $|v_x| \le (2kT/m)^{1/2}$. What is the probability that $|v_x| > (2kT/m)^{1/2}$?

$$\text{If } |v_x| \le \left(\frac{2\,kT}{m}\right)^{1/2}, \text{ then}$$

$$\xi^2 \le \frac{m}{2\,kT}\left(\frac{2\,kT}{m}\right) = 1$$

and

$$\xi_0 \le \sqrt{1} \le 1$$

The probability that the particle has $|v_x| \le \left(\frac{2\,kT}{m}\right)^{1/2}$ is given by:

$$f_{\le} = \frac{2}{\sqrt{\pi}}\int_0^1 e^{-\xi^2}\,d\xi$$
$$f_{\le} = \mathrm{erf}\,(1)$$
$$f_{\le} = 0.8427$$

The probability that the particle would have a velocity $|v_x| > \left(\frac{2\,kT}{m}\right)^{1/2}$ is found from the previous part, since the total probability must be 1:

$$f_{>} = 1 - 0.8427$$
$$f_{>} = 0.1573$$

***P34.9*)** The speed of sound is given by $v_{sound} = \sqrt{\frac{\gamma kT}{m}} = \sqrt{\frac{\gamma RT}{M}}$, where $\gamma = C_P/C_V$.

a) What is the speed of sound in Ne, Kr, and Ar at 1000 K?
b) At what temperature will the speed of sound in Kr equal the speed of sound in Ar at 1000 K?

a) $M_{Ne} = 0.020\ \mathrm{kg\ mol^{-1}}$ $\quad M_{kr} = 0.084\ \mathrm{kg\ mol^{-1}}$ $\quad M_{Ar} = 0.040\ \mathrm{kg\ mol^{-1}}$

The heat capacities for the three gases are the same:

$C_v = 12.5\ \text{J K}^{-1}\text{mol}^{-1}$ and $C_p = 20.8\ \text{J K}^{-1}\text{mol}^{-1}$

Thus, $\gamma = 1.6666$.

a) Ne:

$$v_{\text{sound}} = \sqrt{\frac{1.67\left(8.314\ \text{J mol}^{-1}\text{K}\right)\left(1000\ \text{K}\right)}{0.020\ \text{kg mol}^{-1}}} = 829\ \text{m s}^{-1}$$

Kr:

$$v_{\text{sound}} = \sqrt{\frac{1.67\left(8.314\ \text{J mol}^{-1}\text{K}^{-1}\right)\left(1000\ \text{K}\right)}{0.084\ \text{kg mol}^{-1}}} = 407\ \text{m s}^{-1}$$

Ar:

$$v_{\text{sound}} = \sqrt{\frac{1.67\left(8.314\ \text{J mol}^{-1}\text{K}^{-1}\right)\left(1000\ \text{K}\right)}{0.040\ \text{kg mol}}} = 589\ \text{m s}^{-1}$$

b) Setting the speed of sound equal for Kr and Ar:

$$\sqrt{\frac{\gamma RT_{\text{Kr}}}{M_{\text{Kr}}}} = \sqrt{\frac{\gamma RT_{\text{Ar}}}{M_{\text{Ar}}}}$$

$$T_{\text{Kr}} = \frac{M_{\text{Kr}}}{M_{\text{Ar}}} \cdot T_{\text{Ar}}$$

Therefore:

$$T_{\text{Kr}} = \left(\frac{0.084\ \text{kg mol}^{-1}}{0.040\ \text{kg mol}^{-1}}\right) 1000\ \text{K}$$

$$T_{\text{Kr}} = (2.10)1000\ \text{K}$$

$$T_{\text{Kr}} = 2100\ \text{K}$$

P34.15) Starting with the Maxwell speed distribution, demonstrate that the probability distribution for translational energy for $\varepsilon_{Tr} >> kT$ is given by:

$$f\left(\varepsilon_{Tr}\right) d\varepsilon_{Tr} = 2\pi \left(\frac{1}{\pi kT}\right)^{3/2} e^{-\varepsilon_{Tr}/kT} \varepsilon_{Tr}^{1/2} d\varepsilon_{Tr}$$

The translational energy of a particle can be related to the velocity of the particle by the expressions

$$\varepsilon_{Tr} = \frac{1}{2}\mathrm{m}v^2$$
$$v = \sqrt{\frac{2\varepsilon_T}{\mathrm{m}}}$$
$$\mathrm{d}v = \sqrt{\frac{2}{\mathrm{m}}}\cdot\frac{1}{2}\sqrt{\frac{1}{\varepsilon_{Tr}}}\,\mathrm{d}\varepsilon_{Tr} = \frac{1}{2}\sqrt{\frac{2}{\mathrm{m}\varepsilon_{Tr}}}\,\mathrm{d}\varepsilon_{Tr}$$

Substituting this result into the Maxwell speed distribution:

$$f(\varepsilon_{Tr})\mathrm{d}\varepsilon_{Tr} = 4\pi\left(\frac{m}{2\pi\, kT}\right)^{3/2}\left(\frac{2\varepsilon_{Tr}}{m}\right)\mathrm{e}^{-\frac{m}{2\,kT}\left(\frac{2\varepsilon_{Tr}}{m}\right)}\left(\frac{1}{2}\sqrt{\frac{2}{m\varepsilon_{Tr}}}\,d\varepsilon_{Tr}\right)$$
$$= 2\pi\left(\frac{1}{\pi\mathrm{kT}}\right)^{3/2}\mathrm{e}^{-\frac{\varepsilon_{Tv}}{\mathrm{kT}}}\varepsilon_T^{1/2}\,\mathrm{d}\varepsilon_{Tv}$$

P34.19) Imagine a cubic container with sides 1 cm in length that contains 1 atm of Ar at 298 K. How many gas–wall collisions are there per second?

The collisional rate is given by:

$$\frac{dN_c}{dt} = \frac{PAv_{ave}}{4\,kT} = \frac{PA\,N_A v_{avg\backslash e}}{4RT}$$

With $M_{Ar} = 0.040$ kg mol^{-1} and $T = 298$ K the average speed is:

$$v_{\mathrm{avg}} = \sqrt{\frac{8RT}{\pi M}} = \sqrt{\frac{8(8.314\text{ J mol}^{-1}\text{ K}^{-1})(298\text{ K})}{\pi(0.040\text{ kg mol}^{-1})}}$$
$$v_{\mathrm{avg}} = 459\text{ m s}^{-1}$$

Substituting into the expression for the collisional rate:

$$\frac{\mathrm{dN_c}}{\mathrm{dt}} = \frac{\left(1\text{ atm}\cdot\frac{101.325\times10^3\text{ N m}^{-2}}{1\text{ atm}}\right)\left(1\text{ cm}^2\left(\frac{1\text{ m}}{100\text{ cm}}\right)^2\right)(6.022\times10^{23}\text{ mol}^{-1})(459\text{ m s}^{-1})}{4(8.314\text{ J mol}^{-1}\text{K}^{-1})(298\text{ K})}$$

$$\frac{\mathrm{dN_c}}{\mathrm{dt}} = 2.823\times10^{23}\text{ coll. per sec. per wall}$$

Taking into account the six walls that comprise the container, the total collisional rate is:

$$\frac{dN_c}{dt} = 6 \text{ walls} \cdot 2.823\times10^{23} \text{ coll. s}^{-1} \text{ wall}^{-1}$$
$$\frac{dN_c}{dt} = 1.70\times10^{24} \text{ coll. s}^{-1}$$

P34.21)

a) How many molecules strike a 1-cm^2 surface during 1 min if the surface is exposed to O_2 at 1 atm and 298 K?
b) Ultrahigh vacuum studies typically employ pressures on the order of 10^{-10} Torr. How many collisions will occur at this pressure at 298 K?

For O_2, M = 0.032 kg mol^{-1} and $Z_c = \dfrac{PN_A}{(2\pi MRT)^{1/2}}$

a) @ 1 atm: $$Z_c = \frac{(1 \text{ atm})\left(\dfrac{101.325\times10^3 \text{ Pa}}{1 \text{ atm}}\right)(6.022\times10^{23} \text{ mol}^{-1})}{\left(2\pi(0.032 \text{ kg mol}^{-1})(8.314 \text{ J mol}^{-1}\text{K}^{-1})(298 \text{ K})\right)^{1/2}}$$

$$Z_c = 2.73\times10^{27} \text{ m}^{-2}\text{s}^{-1}$$

$$\frac{dN_c}{dt} = Z_c \times A = (2.73\times10^{27} \text{ m}^{-2}\text{s}^{-1})(1 \text{ cm}^2)\left(\frac{1 \text{ m}}{100 \text{ cm}}\right)^2$$
$$\frac{dN_c}{dt} = 2.73\times10^{23} \text{ coll. s}^{-1}$$

b) @ 10^{-10} torr: $Z_c = \dfrac{\left(10^{-10}\text{ torr}\right)\left(\dfrac{133.32\text{ Pa}}{1\text{ torr}}\right)\left(6.022\times10^{23}\text{ mol}^{-1}\right)}{\left(2\pi\left(0.032\text{ kg mol}^{-1}\right)\left(8.314\text{ J mol}^{-1}\text{K}^{-1}\right)\left(298\text{ K}\right)\right)^{1/2}}$

$$Z_c = 3.60\times10^{14}\text{ m}^{-2}\text{s}^{-1}$$

$$\frac{dN_c}{dt} = Z_c\times A = \left(3.60\times10^{14}\text{ m}^{-2}\text{s}^{-1}\right)\left(1\text{ cm}^2\right)\left(\frac{1\text{ m}}{100\text{ cm}}\right)^2$$

$$\frac{dN_c}{dt} = 3.60\times10^{14}\text{ coll s}^{-1}$$

P34.23) Many of the concepts developed in this chapter can be applied to understanding the atmosphere. Because atmospheric air is comprised primarily of N_2 (roughly 78% by volume), approximate the atmosphere as consisting only of N_2 in answering the following questions:

a) What is the single-particle collisional frequency at sea level, with $T = 298$ K and $P = 1$ atm? The corresponding single-particle collisional frequency is reported as 10^{10} s^{-1} in the *CRC Handbook of Chemistry and Physics* (62nd ed., p. F-171).
b) At the tropopause (11 km in altitude), the collisional frequency decreases to $3.16\times10^9\text{ s}^{-1}$, primarily due to a reduction in temperature and barometric pressure (i.e., fewer particles). The temperature at the tropopause is ~220 K. What is the pressure of N_2 at this altitude?
c) At the tropopause, what is the mean free path for N_2?

The collisional cross section of N_2 is $\sigma = 4.3\times10^{-19}\text{ m}^2$, and $M = 0.028\text{ kg mol}^{-1}$.

a) $z_{11} = \sqrt{2}\sigma\dfrac{PN_A}{RT}\left(\dfrac{8RT}{\pi M}\right)^{1/2}$

$$= \sqrt{2}\left(4.3\times10^{-19}\text{ m}^2\right)\frac{\left(1\text{ atm}\right)\left(6.022\times10^{23}\text{ mol}^{-1}\right)}{\left(8.21\times10^{-2}\text{ L atm mol}^{-1}\text{ K}^{-1}\right)\left(298\text{ K}\right)}\left(\frac{8\left(8.314\text{ J mol}^{-1}\text{ K}^{-1}\right)\left(298\text{ K}\right)}{\pi\left(0.028\text{ kg mol}^{-1}\right)}\right)^{1/2}$$

$$= \sqrt{2}\left(2.46\times10^{22}\text{ L}^{-1}\right)\left(4.3\times10^{-19}\text{ m}^2\right)\left(475\text{ m s}^{-1}\right)\left(\frac{1000\text{ L}}{\text{m}^3}\right)$$

$$z_{11} = 7.11\times10^9\text{ s}^{-1}$$

b) $$3.16\times10^{9}\ \text{s}^{-1} = \frac{\sqrt{2}(\text{P})(6.022\times10^{23}\ \text{mol}^{-1})}{(8.21\times10^{-2}\ \text{L atm mol}^{-1}\ \text{K}^{-1})(220\ \text{K})}(4.3\times10^{-19}\ \text{m}^{2})\left(\frac{8(8.314\ \text{J mol}^{-1}\ \text{K}^{-1})(220\ \text{K})}{\pi(0.028\ \text{kg mol}^{-1})}\right)^{1/2}$$

$$\text{P} = \frac{(3.16\times10^{9}\ \text{s}^{-1})}{\sqrt{2}(3.34\times10^{22}\ \text{L}^{-1}\ \text{atm}^{-1})(4.3\times10^{-19}\ \text{m}^{2})(408\ \text{m s}^{-1})}\left(\frac{1\ \text{m}^{3}}{1000\ \text{L}}\right)$$

$$\text{P} = 0.382\ \text{atm}$$

c) $$\lambda = \left(\frac{RT}{\text{PN}_{\text{A}}}\right)\frac{1}{\sqrt{2}\sigma} = \left(\frac{(8.21\times10^{-2}\ \text{L atm mol}^{-1}\ \text{s}^{-1})(220\ \text{K})}{(0.382\ \text{atm})(6.022\times10^{23}\ \text{mol}^{-1})}\right)\frac{1}{\sqrt{2}(4.3\times10^{-19}\ \text{m}^{2})}$$

$$\lambda = \left(\frac{18.1\ \text{L atm mol}^{-1}}{1.40\times10^{5}\ \text{atm mol}^{-1}\ \text{m}^{2}}\right)\left(\frac{1\ \text{m}^{3}}{1000\ \text{L}}\right)$$

$$\lambda = 1.29\times10^{-7}\ \text{m}$$

P34.26) Determine the mean free path for Ar at 298 K at the following pressures:
a) 0.5 atm
b) 0.005 atm
c) 5×10^{-6} atm

For Ar, $\sigma = 3.6\times10^{-19}\ \text{m}^2$ and $\text{M} = 0.040\ \text{kg mol}^{-1}$

a) $$\lambda = \left(\frac{RT}{\text{PN}_{\text{A}}}\right)\frac{1}{\sqrt{2}\sigma} = \left(\frac{(8.21\times10^{-2}\ \text{L atm mol}^{-1}\ \text{K}^{-1})(298\ \text{K})}{(0.5\ \text{atm})(6.022\times10^{23}\ \text{mol}^{-1})}\right)\frac{1}{\sqrt{2}\ (3.6\times10^{-19}\ \text{m}^{2})}\left(\frac{1\ \text{m}^{3}}{1000\ \text{L}}\right)$$

$$\lambda = 1.60\times10^{-7}\ \text{m}$$

b) Since the mean free path is inversely proportional to pressure, the result from part (a) can be used to determine the mean free path at pressures specified in parts (b) and (c) as follows:

$$\lambda_{0.005} = \lambda_{0.5}\left(\frac{0.5\ \text{atm}}{0.005\ \text{atm}}\right) = 1.60\times10^{-7}\ \text{m}\ (100)$$

$$\lambda_{0.005} = 1.60\times10^{-5}\ \text{m}$$

c) $$\lambda_{5\times10^{-6}} = \lambda_{0.5}\left(\frac{0.5\ \text{atm}}{5\times10^{-6}\ \text{atm}}\right) = 1.60\times10^{-7}\ \text{m}\ (10^{5})$$

$$\lambda_{5\times10^{-6}} = 1.60\times10^{-2}\ \text{m}$$

P34.29) A comparison of v_{ave}, v_{mp}, and v_{rms} for the Maxwell speed distribution reveals that these three quantities are not equal. Is the same true for the one-dimensional velocity distributions?

$$v_{avg} = \langle v \rangle = \int_{-\infty}^{\infty} v_x \left(\frac{M}{2\pi RT}\right)^{1/2} e^{-\frac{M}{2RT}v_x^2}\, dv_x$$

$$= \left(\frac{M}{2\pi RT}\right)^{1/2} \int_{-\infty}^{\infty} v_x e^{-\frac{M}{2RT}v_x^2}\, dv_x$$

$$v_{avg} = 0$$

$$v_{mp} \Rightarrow 0 = \frac{\partial}{\partial v_x}\left[\left(\frac{M}{2\pi RT}\right)^{1/2} e^{-\frac{M}{2RT}v_x^2}\right]$$

$$0 = \left(\frac{M}{2\pi RT}\right)^{1/2}\left(\frac{M}{2\pi RT}v_x\right)e^{-\frac{M}{2RT}v_x^2}$$

The above equality will be true when $v_x = 0$; therefore, $v_{mp} = 0$.

$$v_{rms} = \langle v^2 \rangle^{1/2} = \left[\int_{-\infty}^{\infty} v_x^2 \left(\frac{M}{2\pi RT}\right)^{1/2} e^{-\frac{M}{2RT}v_x^2}\, dv_x\right]^{1/2}$$

$$= \left[\left(\frac{\beta}{\pi}\right)^{1/2} \int_{-\infty}^{\infty} v_x^2 e^{-\beta v_x^2}\, dv_x\right]^{1/2} \quad \text{for } \beta = \frac{M}{2RT}$$

$$= \left[\left(\frac{\beta}{\pi}\right)^{1/2} \left(2\int_{0}^{\infty} v_x^2 e^{-\beta v_x^2}\, dv_x\right)\right]^{1/2}$$

$$= \left[\left(\frac{\beta}{\pi}\right)^{1/2} \cdot \frac{1}{2}\left(\frac{\pi}{\beta^3}\right)^{1/2}\right]^{1/2}$$

$$= \left[\frac{1}{2}\beta^{-1}\right]^{1/2}$$

$$v_{rms} = \sqrt{\frac{RT}{M}}$$

Chapter 35: Transport Phenomena

P35.2)

a) The diffusion coefficient for Xe at 273 K and 1 atm is 0.5×10^{-5} m^2 s^{-1}. What is the collisional cross section of Xe?

b) The diffusion coefficient of N_2 is threefold greater than that of Xe under the same pressure and temperature conditions. What is the collisional cross section of N_2?

a)

$$D_{Xe} = 0.5\times10^{-5}\,\text{m}^2\text{s}^{-1} \quad @\ 273\ \text{K and 1 atm}$$

$$\sigma = \frac{1}{3}\sqrt{\frac{8kT}{\pi M}}\left(\frac{RT}{PN_A}\right)\frac{1}{\sqrt{2}D}$$

$$= \frac{1}{3\sqrt{2}}\sqrt{\frac{8\left(8.314\ \text{J mol}^{-1}\ \text{K}^{-1}\right)\left(273\ \text{K}\right)}{\pi\left(0.131\ \text{kg mol}^{-1}\right)}}\left(\frac{\left(8.21\times10^{-2}\ \text{L atm mol}^{-1}\ \text{K}\right)\left(\dfrac{1\ \text{m}^3}{1000\ \text{L}}\right)\left(273\ \text{K}\right)}{\left(1\ \text{atm}\right)\left(6.022\times10^{23}\ \text{mol}^{-1}\right)\left(0.5\times10^{-5}\ \text{m}^2\ \text{s}^{-1}\right)}\right)$$

$$\sigma = 0.368\ \text{nm}^2$$

b) The ratio of collisional cross sections is given by:

$$\frac{\sigma_{N_2}}{\sigma_{Xe}} = \frac{D_{Xe}}{D_{N_2}}\sqrt{\frac{M_{Xe}}{M_{N_2}}}$$

$$\sigma_{N_2} = \sigma_{Xe}\left(\frac{D_{Xe}}{D_{N_2}}\right)\sqrt{\frac{M_{Xe}}{M_{N_2}}}$$

$$= \left(0.368\ \text{nm}^2\right)\left(\frac{1}{3}\right)\sqrt{\frac{0.131\ \text{kg mol}^{-1}}{0.028\ \text{kg mol}^{-1}}}$$

$$\sigma_{N_2} = 0.265\ \text{nm}^2$$

P35.6) A thermopane window consists of two sheets of glass separated by a volume filled with air (which we will model as N_2 where $\kappa = 0.0240$ J K^{-1} m^{-1} s^{-1}). For a thermopane window that is 1 m^2 in area with a separation between glass sheets of 3 cm, what is the loss of energy when:

a) the exterior of the window is at a temperature of 10°C and the interior of the window is at a temperature of 22°C?

b) the exterior of the window is at a temperature of –20°C and the interior of the window is at a temperature of 22°C?
c) the same temperature differential as in part (b) is used but the window is filled with Ar ($\kappa = 0.0163 \text{ J K}^{-1} \text{ m}^{-1} \text{ s}^{-1}$) rather than N_2?

The energy flux is given by

$$J = -\kappa\left(\frac{dT}{dx}\right) = -\kappa\left(\frac{\Delta T}{\Delta x}\right)$$

The loss in energy is equal to the flux times the area (A) through which the energy loss occurs:

$$\Delta E = -\kappa\left(\frac{\Delta T}{\Delta x}\right)\cdot A$$

a) $$\Delta E = -\left(0.0240 \text{ J K}^{-1} \text{ m}^{-1} \text{ s}^{-1}\right)\left(\frac{12 \text{ K}}{3 \text{ cm}}\right)\left(\frac{100 \text{ cm}}{1 \text{ m}}\right)\left(1 \text{ m}^2\right)$$

$$= -9.60 \text{ J s}^{-1}$$

b) $\Delta T = 42 \text{ K}$

$$\Delta E = -\left(0.0240 \text{ J K}^{-1} \text{ m}^{-1} \text{ s}^{-1}\right)\left(\frac{42 \text{ K}}{3 \text{ cm}}\right)\left(\frac{100 \text{ cm}}{1 \text{ m}}\right)\left(1 \text{ m}^2\right)$$

$$= -33.6 \text{ J s}^{-1}$$

c) $\Delta T = 42 \text{ K} \qquad \kappa = 0.0163 \text{ J K}^{-1}\text{m}^{-1}\text{s}^{-1}$

$$\Delta E = -\left(0.0163 \text{ J K}^{-1}\text{m}^{-1}\text{s}^{-1}\right)\left(\frac{42 \text{ K}}{3 \text{ cm}}\right)\left(\frac{100 \text{ cm}}{1 \text{ m}}\right)\left(1 \text{ m}^2\right)$$

$$= -22.8 \text{ J s}^{-1}$$

P35.8) Determine the thermal conductivity of the following species at 273 K and 1 atm:
a) Ar ($\sigma = 0.36 \text{ nm}^2$)
b) Cl_2 ($\sigma = 0.93 \text{ nm}^2$)
c) SO_2 ($\sigma = 0.58 \text{ nm}^2$, geometry: bent)
You will need to determine $C_{V,m}$ for the species listed. You can assume that the translational and rotational degrees of freedom are in the high-temperature limit, and that the vibrational contribution to $C_{V,m}$ can be ignored at this temperature.

$$\kappa = \frac{1}{3}\frac{C_{vm}}{N_A}\cdot\left(\frac{8RT}{\pi M}\right)^{1/2}\frac{1}{\sqrt{2}\sigma}$$

a) $C_{V,m}^{Av} = \frac{3}{2}R$

$$\kappa = \frac{1}{3}\left(\frac{3}{2}\frac{R}{N_A}\right)\cdot\left(\frac{8RT}{\pi M}\right)^{1/2}\frac{1}{\sqrt{2}\sigma}$$

$$= \frac{8.314\ \text{J mol}^{-1}\text{K}^{-1}}{2\left(6.022\times10^{23}\ \text{mol}^{-1}\right)}\cdot\left(\frac{8\left(8.314\ \text{J mol}^{-1}\ \text{K}^{-1}\right)\left(273\ \text{K}\right)}{\pi\left(0.040\ \text{kg mol}^{-1}\right)}\right)^{1/2}\frac{1}{\sqrt{2}\left(0.36\ \text{nm}^2\right)}\cdot\left(\frac{10^9\ \text{nm}}{1\ \text{m}}\right)^2$$

$$= 0.00516\ \text{J K}^{-1}\ \text{m}^{-1}\ \text{s}^{-1}$$

b) $C_{V,m}^{Cl_2} = C_V^T + C_V^R = \frac{3}{2}R + R = \frac{5}{2}R$

$$\kappa = \frac{1}{3}\left(\frac{5}{2}\frac{R}{N_A}\right)\left(\frac{8RT}{\pi M}\right)^{1/2}\frac{1}{\sqrt{2}\sigma}$$

$$= \frac{5\left(8.314\ \text{J mol}^{-1}\ \text{K}^{-1}\right)}{6\left(6.022\times10^{23}\ \text{mol}^{-1}\right)}\left(\frac{8\left(8.314\ \text{J mol}^{-1}\ \text{K}^{-1}\right)\left(273\ \text{K}\right)}{\pi\left(0.071\ \text{kg mol}^{-1}\right)}\right)^{1/2}\frac{1}{\sqrt{2}\left(0.93\ \text{nm}^2\right)}\left(\frac{10^9\ \text{nm}}{1\ \text{m}}\right)^2$$

$$= 0.00249\ \text{J K}^{-1}\ \text{m}^{-1}\ \text{s}^{-1}$$

c) $C_{V,m}^{SO_2} = C_V^T + C_V^R = \frac{3}{2}R + \frac{3}{2}R = 3R$

$$\kappa = \frac{1}{3}\left(\frac{3R}{N_A}\right)\left(\frac{8RT}{\pi M}\right)^{1/2}\frac{1}{\sqrt{2}\sigma}$$

$$= \frac{\left(8.314\ J\,mol^{-1}\ K^{-1}\right)}{3\left(6.022\times10^{23}\ mol^{-1}\right)}\left(\frac{8\left(8.314\ J\,mol^{-1}\ K^{-1}\right)\left(273\ K\right)}{\pi\left(0.064\ kg\ mol^{-1}\right)}\right)^{1/2}\frac{1}{\sqrt{2}\left(0.58\ nm^2\right)}\left(\frac{10^9\ nm}{1\ m}\right)^2$$

$$= 0.0050\ J\,K^{-1}\ m^{-1}\ s^{-1}$$

P35.9) The thermal conductivity of Kr is 0.0087 J K^{-1} m^{-1} s^{-1} at 273 K and 1 atm. Estimate the collisional cross section of Kr.

Treating Kr as an ideal monatomic gas, $C_{V,m} = \frac{3}{2}R$, and the thermal conductivity is:

$$\kappa = \frac{1}{3}C_{V,m}v_{ave}\frac{1}{\sqrt{2}\sigma} = \frac{1}{3}\left(\frac{3}{2}\frac{R}{N_A}\right)\left(\frac{8RT}{\pi M}\right)^{1/2}\frac{1}{\sqrt{2}\sigma}$$

Rearranging to isolate the collisional cross section:

$$\sigma = \frac{1}{3} C_{V,m} v_{ave} \frac{1}{\sqrt{2}\sigma} = \frac{1}{3}\left(\frac{3}{2}\frac{R}{N_A}\right)\left(\frac{8RT}{\pi M}\right)^{1/2} \frac{1}{\sqrt{2}\kappa}$$

$$\sigma = \frac{\left(8.314\ J\ mol^{-1}\ K^{-1}\right)}{2\left(6.022\times 10^{23}\ mol^{-1}\right)}\left(\frac{8\left(8.314\ J\ mol^{-1}\ K^{-1}\right)\left(273\ K\right)}{\pi\left(0.084\ kg\ mol^{-1}\right)}\right)^{1/2} \frac{1}{\sqrt{2}\left(0.0087\ J K^{-1}\ m^{-1}\ s^{-1}\right)}$$

$$= 1.5\times 10^{-19}\ m^2$$

P35.12) The thermal conductivities of acetylene (C_2H_2) and N_2 at 273 K and 1 atm are 0.01866 and 0.0240 J m^{-1} s^{-1} K^{-1}, respectively. Based on these data, what is the ratio of the collisional cross section of acetylene relative to N_2?

$$\frac{\kappa_{C_2H_2}}{\kappa_{N_2}} = \frac{\frac{C_{V,m}^{C_2H_2}}{3N_A}\left(\frac{8RT}{\pi M_{C_2H_2}}\right)^{1/2}\frac{1}{\sqrt{2}\sigma_{C_2H_2}}}{\frac{C_{V,m}^{N_2}}{3N_A}\left(\frac{8RT}{\pi M_{N_2}}\right)^{1/2}\frac{1}{\sqrt{2}\sigma_{N_2}}}$$

$$= \frac{C_{V,m}^{C_2H_2}}{C_{V,m}^{N_2}}\left(\frac{M_{N_2}}{M_{C_2H_2}}\right)^{1/2}\frac{\sigma_{N_2}}{\sigma_{C_2H_2}}$$

Rearranging to isolate the ratio of collisional cross sections:

$$\frac{\sigma_{C_2H_2}}{\sigma_{N_2}} = \frac{C_{V,m}^{C_2H_2}}{C_{V,m}^{N_2}}\left(\frac{M_{N_2}}{M_{C_2H_2}}\right)^{1/2}\frac{\kappa_{N_2}}{\kappa_{C_2H_2}}$$

Both C_2H_2 and N_2 are linear molecules, and will therefore have the same heat capacity value so that the collision cross section ratio depends only on the mass and thermal conductivity ratios:

$$\frac{\sigma_{C_2H_2}}{\sigma_{N_2}} = \left(\frac{0.028\ \text{kg mol}^{-1}}{0.026\ \text{kg mol}^{-1}}\right)^{1/2}\left(\frac{0.0240\ \text{J m}^{-1}\ \text{s}^{-1}\ \text{K}^{-1}}{0.01866\ \text{J m}^{-1}\ \text{s}^{-1}\ \text{K}^{-1}}\right)$$

$$\frac{\sigma_{C_2H_2}}{\sigma_{N_2}} = 1.33$$

P35.16) The viscosity of H_2 at 273 K at 1 atm is 84 μP. Determine the viscosities of D_2 and HD.

The expression for viscosity is:

$$\eta = \frac{1}{3}\left(\frac{8RT}{\pi M}\right)^{1/2}\frac{1}{\sqrt{2}\sigma}\frac{M}{N_A}$$

Taking the ratio of viscosities for two species (denoted as 1 and 2) yields

$$\frac{\eta_2}{\eta_1} = \sqrt{\frac{M_2}{M_1}}\left(\frac{\sigma_1}{\sigma_2}\right)$$

Assuming that the collisional cross sections for the species are the same, the ratio of velocities reduces to:

$$\frac{\eta_2}{\eta_1} = \sqrt{\frac{M_2}{M_1}}$$

Substituting the molecular weights into the above expression yields the following viscosities for D_2 and HD:

$$\eta_{D_2} = \eta_{H_2}\sqrt{\frac{M_{D_2}}{M_{H_2}}}$$

$$= (84\ \mu\text{P})\sqrt{\frac{4.04\ \text{g mol}^{-1}}{2.02\ \text{g mol}^{-1}}}$$

$$\eta_{D_2} = 118\ \mu\text{P}$$

$$\eta_{HD} = 84\ \mu\text{P}\sqrt{\frac{3.03\ \text{g mol}^{-1}}{2.02\ \text{g mol}^{-1}}}$$

$$= 103\ \mu\text{P}$$

P35.18) How long will it take to pass 200 mL of H_2 at 273 K through a 10-cm-long capillary tube of 0.25 mm if the gas input and output pressures are 1.05 and 1.00 atm, respectively?

The flow rate is given as

$$\frac{\Delta V}{\Delta t} = \frac{\pi r^4}{8\eta}\left(\frac{P_2 - P_1}{x_2 - x_1}\right)$$

Substituting into this expression and solving for Δt yields:

$$\frac{(0.2 \text{ L})}{\Delta t} = \frac{\pi\left(2.5\times10^{-4} \text{ m}\right)^4}{8\left(84\times10^{-6}\text{P}\right)\left(\dfrac{0.1 \text{ kg m}^{-1}\text{s}^{-1}}{1 \text{ P}}\right)}\left(\frac{1.05 \text{ atm} - 1.00 \text{ atm}}{0.1 \text{ m}}\right)$$

$$\frac{(0.2 \text{ L})}{\Delta t} = 9.13\times10^{-11} \text{ m}^4 \text{ kg}^{-1} \text{ s} \cdot \frac{101{,}325 \text{ N m}^{-2}}{1 \text{ atm}}.$$

$$\frac{(0.2 \text{ L})}{\Delta t} = 9.13\times10^{-6} \text{ m}^3 \text{ s}^{-1}$$

$$\frac{(0.2 \text{ L})\left(\dfrac{1 \text{ m}^3}{1000 \text{ L}}\right)}{9.13\times10^{-5} \text{ m}^3 \text{ s}^{-1}} = 21.9 \text{ s} = \Delta t$$

P35.22) Myoglobin is a protein that participates in oxygen transport. For myoglobin in water at 20°C, $\overline{s} = 2.04 \times 10^{-13}$ s, $D = 1.13 \times 10^{-10}$ m^2 s^{-1}, and $\overline{V} = 0.740$ cm^3 g^{-1}. The density of water is 0.998 g cm^3 and the viscosity is 1.002 cP at this temperature.

a) Using the information provided, estimate the size of myoglobin.
b) What is the molecular weight of myoglobin?

a) Using the Stokes-Einstein equation, the radius of myoglobin is:

$$r = \frac{kT}{6\pi\eta D}$$

$$r = \frac{kT}{6\pi\eta D}$$

$$= \frac{\left(1.38\times10^{-23} \text{ J K}^{-1}\right)(293 \text{ K})}{6\pi(0.01002 \text{ P})\left(\dfrac{0.1 \text{ kg m}^{-1}\text{s}^{-1}}{1 \text{ P}}\right)\left(1.13\times10^{-10} \text{ m}^2 \text{ s}^{-1}\right)}$$

$$= 1.89\times10^{-9} \text{ m}$$

$$= 1.89 \text{ nm}$$

b) The molecular weight of myoglobin can be found as follows:

$$M = \frac{RT\overline{s}}{D\left(1-\overline{V}\rho\right)}$$

$$= \frac{\left(8.314 \text{ J mol}^{-1} \text{ K}^{-1}\right)(293 \text{ K})\left(2.04\times10^{-13} \text{ s}\right)}{\left(1.13\times10^{-10} \text{ m}^2 \text{ s}^{-1}\right)\left(1-\left(0.740 \text{ cm g}^{-1}\right)\left(0.998 \text{ g cm}^{-3}\right)\right)}$$

$$= 16.8 \text{ kg mol}^{-1}$$

P35.24) Boundary centrifugation is performed at an angular velocity of 40,000 rpm to determine the sedimentation coefficient of cytochrome *c* (M = 13,400 g mol^{-1}) in water at

20°C ($\rho = 0.998$ g cm^{-3}, $\eta = 1.002$ cP). The following data are obtained on the position of the boundary layer as a function of time:

Time (h)	x_b (cm)
0	4.00
2.5	4.11
5.2	4.23
12.3	4.57
19.1	4.91

a) What is the sedimentation coefficient for cytochrome *c* under these conditions?
b) The specific volume of cytochrome *c* is 0.728 cm^3 g^{-1}. Estimate the size of cytochrome *c*.

a) Using the data from the table, a plot of $ln\left(\frac{x_b}{x_{b,t=0}}\right)$ versus t can be constructed, the slope of which is equal to $\omega^2\overline{s}$:

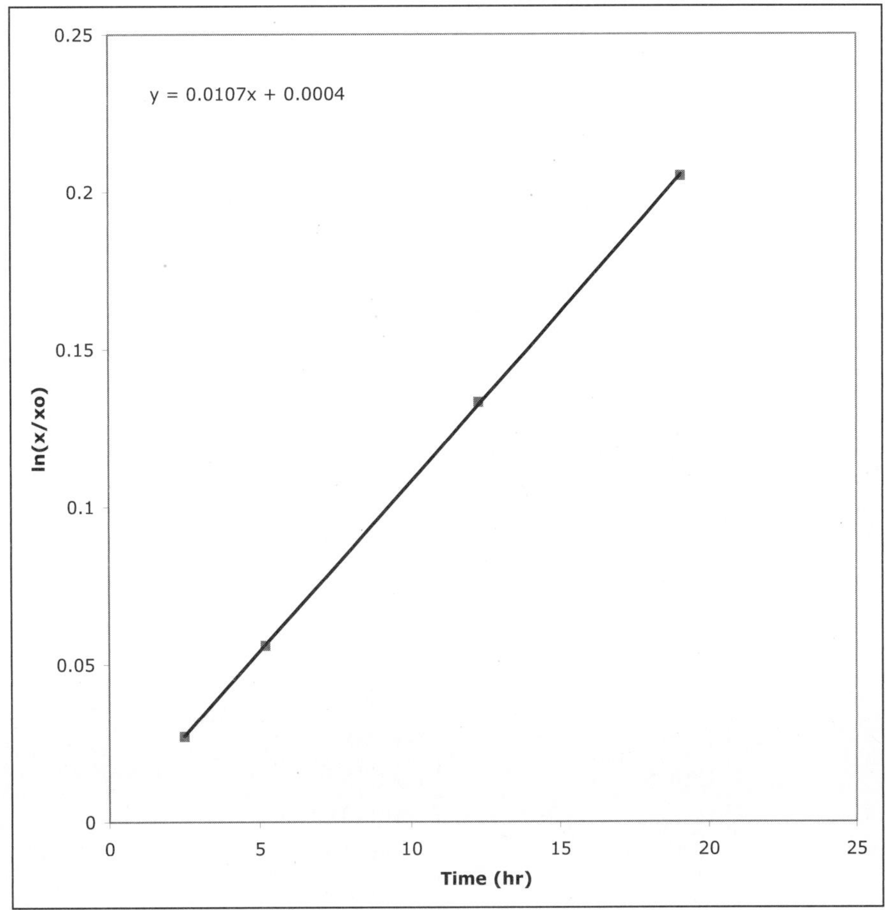

The slope from the best fit to the line is 0.0107 hr^{-1}. Using this slope, the sedimentation coefficient is determined as follows:

$$\omega^2\bar{s} = 0.0107\ \text{hr}^{-1}$$

$$\bar{s} = \frac{0.0107\ \text{hr}^{-1}}{\omega^2}\left(\frac{1\ \text{hr}}{3600\ \text{s}}\right) = \frac{2.97\times10^{-6}\ \text{s}^{-1}}{\omega^2}$$

$$= \frac{2.97\times10^{-6}\ \text{s}^{-1}}{\left(4.00\times10^{4}\ \text{rev min}^{-1}\right)^2\left(\frac{2\pi\ \text{rad}}{1\ \text{rev}}\right)^2\left(\frac{1\ \text{min}}{60\ \text{sec}}\right)^2}$$

$$= 1.70\times10^{-13}\ \text{s}$$

b) First, the frictional force is calculated as follows:

$$f = \frac{\text{m}\left(1-\bar{V}\rho\right)}{\bar{s}}$$

$$= \frac{\left(\frac{13.4\ \text{kg mol}^{-1}}{6.022\times10^{23}\ \text{mol}^{-1}}\right)\left(1-\left(0.728\ \text{cm}^3\text{g}^{-1}\right)\left(0.998\ \text{g cm}^{-3}\right)\right)}{1.70\times10^{-13}\ \text{s}}$$

$$f = 3.58\times10^{-11}\ \text{kg s}^{-1}$$

With the frictional force, the particle radius can be determined:

$$6\pi\eta r = f$$

$$r = \frac{f}{6\pi\eta} = \frac{3.58\times10^{-11}\ \text{kg s}^{-1}}{6\pi\left(0.01005\ \text{P}\right)\left(\frac{0.1\ \text{kg m}^{-1}\ \text{s}^{-1}}{\text{P}}\right)}$$

$$= 1.89\times10^{-9}\ \text{m}$$

$$= 1.89\ \text{nm}$$

P35.28) The molar conductivity of sodium acetate, CH_3COONa, is measured as a function of concentration in water at 298 K, and the following data are obtained:

Concentration (M)	Λ_m (S m^2 mol^{-1})
0.0005	0.00892
0.001	0.00885
0.005	0.00857
0.01	0.00838
0.02	0.00812
0.05	0.00769
0.1	0.00728

Is sodium acetate a weak or strong electrolyte? Determine Λ_m^0 using appropriate methodology depending on your answer.

If the electrolyte is strong, a plot of Λ_m versus $\sqrt{\frac{c}{c_0}}$ (assuming c_0 = 1 M) should yield a straight line. The corresponding plot is shown below:

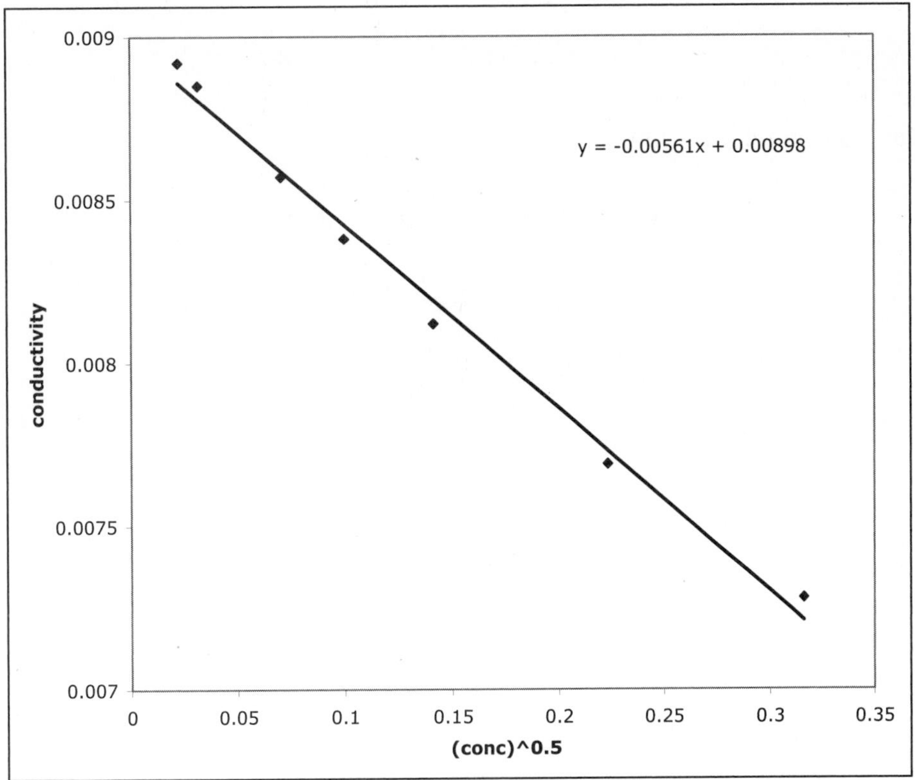

The linearity of the plot demonstrates that sodium acetate is a strong electrolyte. Best fit by a straight line yields the following:

$$\Lambda_m = -0.00561 \cdot \sqrt{\frac{c}{c_0}} + 0.00898$$

Therefore, for sodium acetate $\Lambda_m^0 = 0.00898$ S m^2 mol^{-1}.

P35.32) In the determination of molar conductivities, it is convenient to define the cell constant, K, as $K = \frac{l}{A}$, where l is the separation between the electrodes in the conductivity cell, and A is the area of the electrodes.

a) A standard solution of KCl (conductivity or κ = 1.06296 × 10^{-6} S m^{-1} at 298 K) is employed to standardize the cell, and a resistance of 4.2156 Ω is measured. What is the cell constant?

b) The same cell is filled with a solution of HCl and a resistance of 1.0326 Ω is measured.
What is the conductivity of the HCl solution?

a) The conductivity is defined as

$$\kappa = \frac{\ell}{R \cdot A} = \frac{K}{R}$$

where R is the resistance and K is the cell constant. Using this relationship, the cell constant is determined as follows:

$$\kappa = 1.06296 \times 10^{-6} \text{ s m}^{-1} \qquad \text{R} = 4.2156\Omega$$

$$K = \kappa R = \left(1.06296 \times 10^{-6} \text{ S m}^{-1}\right)\left(4.2156 \ \Omega\right)$$
$$K = 4.48 \times 10^{-6} \text{ S m}^{-1} \ \Omega$$

b)

$$\kappa = \frac{K}{R} = \frac{4.48 \times 10^{-6} \text{ S m}^{-1} \ \Omega}{1.0326 \ \Omega}$$
$$= 4.34 \times 10^{-6} \text{ S m}^{-1}$$

Chapter 36: Elementary Chemical Kinetics

P36.2) Consider the first-order decomposition of cyclobutane at 438°C at constant volume:
$C_4H_8(g) \longrightarrow 2C_2H_4(g)$

a) Express the rate of the reaction in terms of the change in total pressure as a function of time.
b) The rate constant for the reaction is $2.48 \times 10^{-4}\ s^{-1}$. What is the half-life?
c) After initiation of the reaction, how long will it take for the initial pressure of C_4H_8 to drop to 90% of its initial value?

a) The rate is given as

$$R = -\frac{1}{RT}\frac{d\,P_{C_4H_8}}{dt} = \frac{1}{2RT}\frac{d\,P_{C_2H_4}}{dt}$$

The pressure at time *t* is given as $P_t = P_{t=0} - P_{C_4H_8} + P_{C_2H_4}$. Where $P_{t=0}$ is the initial pressure, and $P_{C_4H_8}$ and $P_{C_2H_4}$ represent the pressures of the individual gases as a specific time. By stoichiometry $P_{C_4H_8} = \frac{1}{2}P_{C_2H_4}$ so that:

$$P_t = P_{t=0} - P_{C_2H_4} + 2P_{C_2H_4} = P_{t=0} + P_{C_2H_4}$$

Since $P_{C_2H_4}$ is dependent on time and the initial pressure is time independent:

$$\frac{dP_t}{dt} = \frac{d\,P_{C_2H_4}}{dt}$$

and

$$R = \frac{1}{2RT}\frac{dP_t}{dt}$$

b) The reaction is first order with respect to

$$t_{1/2} = \frac{ln2}{k}$$

so

$$t_{1/2} = \frac{0.693}{2.48\times10^{-4}\ s^{-1}}$$
$$= 2.79\ \times10^{3}s$$

c) Using the integrated rate law

$$-kt = ln\left(\frac{[\mathrm{A}]}{[\mathrm{A}]_0}\right)$$

where

$$\frac{[\mathrm{A}]}{[\mathrm{A}]_0} = 0.90$$

one finds

$$\begin{aligned} t &= -\frac{ln(0.90)}{2.48\times10^{-4}\ \mathrm{s}^{-1}} \\ &= \frac{0.105}{2.48\times10^{-4}\ \mathrm{s}^{-1}} \\ &= 425\ \mathrm{s} \end{aligned}$$

P36.4) Consider the following reaction involving bromophenol blue (BPB) and OH: $\mathrm{BPB}(aq) + \mathrm{OH}^-(aq) \longrightarrow \mathrm{BPBOH}^-(aq)$.
The concentration of BPB can be monitored by following the absorption of this species and using the Beer–Lambert law. In this law, absorption, A, and concentration are linearly related.

a) Express the reaction rate in terms of the change in absorbance as a function of time.
b) Let A_o be the absorbance due to BPB at the beginning of the reaction. Assuming that the reaction is first order with respect to both reactants, how is the absorbance of BPB expected to change with time?
c) Given your answer to part (b), what plot would you construct to determine the rate constant for the reaction?

a) Beer's Law states that

$$\mathrm{A} = \varepsilon \mathrm{b}[\mathrm{BPB}]$$

where ε is the molar absorptivity of BPB and b is the cell path length. The rate of reaction is

$$Rate = -\frac{d[\mathrm{BPB}]}{dt}$$

Thus

$$Rate = -\frac{d[\mathrm{BPB}]}{dt}$$

$$= -\frac{d\left\{\dfrac{\mathrm{A}}{\varepsilon \mathrm{b}}\right\}}{dt}$$

$$Rate = -\frac{1}{\varepsilon \mathrm{b}}\frac{d\mathrm{A}}{dt}$$

b) Let A_0 be the initial absorbance of BPB and A(t) be the absorbance at time, I. Since the reaction is a second order reaction of type II, the integrate rate equation has the form

$$\frac{1}{[\mathrm{OH}^-]_0 - [\mathrm{BPB}]_0}\ln\left(\frac{[\mathrm{OH}^-]/[\mathrm{OH}^-]_0}{[\mathrm{BPB}]/[\mathrm{BPB}]_0}\right) = kt$$

Substituting into the above expression for [BPB] yields:

$$\frac{1}{[\mathrm{OH}^-]_0 - \dfrac{\mathrm{A}_0}{\mathrm{b}\varepsilon}}ln\left(\frac{[\mathrm{OH}^-]/[\mathrm{OH}^-]_0}{\mathrm{A}/\mathrm{A}_0}\right) = kt$$

$$ln\left(\frac{\mathrm{A}}{\mathrm{A}_0}\right) = ln\left(\frac{[\mathrm{OH}^-]}{[\mathrm{OH}^-]_0}\right) - \left([\mathrm{OH}^-] - \frac{\mathrm{A}_0}{b\varepsilon}\right)kt$$

c) A plot of $\ln\left(\dfrac{\mathrm{A}}{\mathrm{A}_0}\right)$ vs. t is predicted to be linear.

P36.8) (Challenging) The first-order thermal decomposition of chlorocyclohexane is as follows: $C_6H_{11}Cl(g) \longrightarrow C_6H_{10}(g) + HCl(g)$. For a constant volume system the following total pressure was measured as a function of time:

Time (s)	P (torr)	Time (min)	P (torr)
3	237.2	24	332.1
6	255.3	27	341.1
9	271.3	30	349.3

12	285.8	33	356.9
15	299.0	36	363.7
18	311.2	39	369.9
21	322.2	42	375.5

a) Derive the following relationship for a first-order reaction: $P(t_2)-P(t_1)=\left(P(t_\infty)-P\left(t_0\right)\right)e^{-kt_1}\left(1-e^{-k(t_2-t_1)}\right)$. In this relation, $P(t_1)$ and $P(t_2)$ are the pressures at two specific times; $P(t_0)$ is the initial pressure when the reaction is initiated, $P(t_\infty)$ is the pressure at the completion of the reaction, and k is the rate constant for the reaction. To derive this relationship:

i. Given the first-order dependence of the reaction, write down the expression for the pressure of chlorocyclohexane at a specific time t_1.

ii. Write the expression for the pressure at another time t_2, which is equal to $t_1 + \Delta$ where delta is a fixed quantity of time.

iii. Write down expressions for $P(t_\infty) - P(t_1)$ and $P(t_\infty) - P(t_2)$.

iv. Subtract the two expressions from part (iii).

b) Using the natural log of the relationship from part (a) and the data provided in the table given earlier in this problem, determine the rate constant for the decomposition of chlorocyclohexane. (*Hint:* Transform the data in the table by defining $t_2 - t_1$ to be a constant value, for example, 9 s.)

a) Since the reaction is first order, we can write:

$$P_\infty - P_{t_1} = \left(P_\infty - P_0\right)e^{-kt_1}$$
$$P_\infty - P_{t_2} = \left(P_\infty - P_0\right)e^{-kt_2} = \left(P_\infty - P_0\right)e^{-k(t_1+\Delta)}$$

subtracting the two previous equations:

$$P_{t_2} - P_{t_1} = \left(P_\infty - P_0\right)e^{-kt_1}\left(1-e^{-k\Delta}\right)$$

where Δ is the difference in time between t_2 and t_1.

b) Taking the natural log of the previous expression yields:

$$\ln\left(P_{t_2} - P_{t_1}\right) = \ln\left(P_\infty - P_0\right) - kt_1 + \ln\left(1-e^{-k\Delta}\right)$$
$$= \ln\left[\left(P_\infty - P_0\right)\left(1-e^{-k\Delta}\right)\right] - kt_1$$

Therefore, a plot of the difference in pressure at fixed difference in time versus t_1 should yield a straight line with slope equal to – k. Using a difference in time of 9 s yields the following table of the difference in pressures versus time:

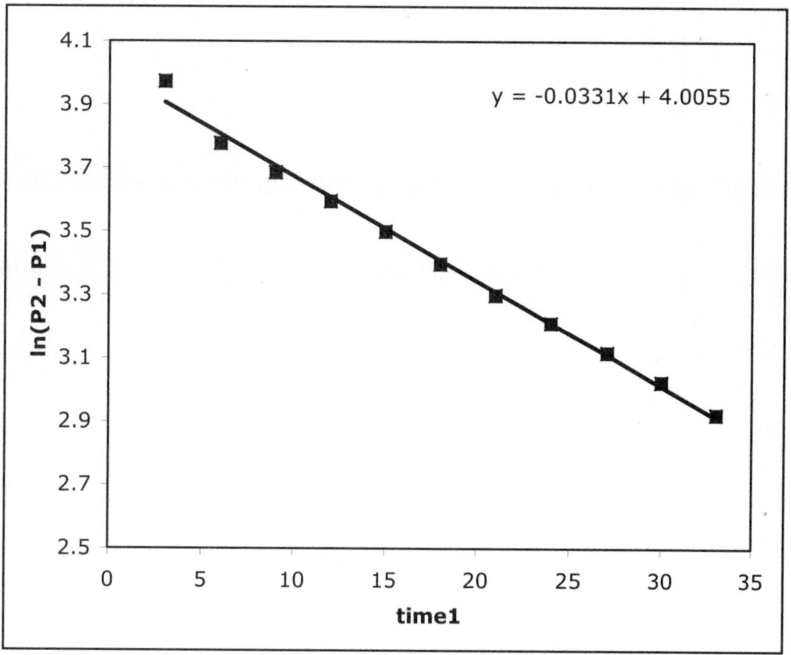

The slope of the best fit line is –0.0311 min^{-1}; therefore, the rate constant is 0.0311 min^{-1}, or 5.18 × 10^{-4} s^{-1}.

P36.12) The half-life of ^{238}U is 4.5 × 10^{9} years. How many disintegrations occur in 1 minute for a 10-mg sample of this element?

The rate constant can be found from the half-life:

$$t_{1/2} = \frac{ln2}{k} \quad or \quad k = \frac{ln2}{t_{1/2}}$$

Thus

$$k = \frac{0.693}{4.5\times10^{9}\ \text{yrs}}\left(\frac{1\ \text{yr}}{364.25\ \text{day}}\right)\left(\frac{1\ \text{day}}{24\ \text{hrs}}\right)\left(\frac{1\ \text{hr}}{60\ \text{min}}\right)$$

$$k = 2.94\times10^{-16}\ \text{min}^{-1}$$

Converting 10 mgs into the number of ^{238}U atoms:

$$N_{^{238}U} = \left(1\times10^{-2}\ \text{g}\right)\left(238\ \text{g mol}^{-1}\right)\left(6.022\times10^{23}\ \text{mol}^{-1}\right) = 1.43\times10^{24}$$

Employing the rate law, the number of disintegrations in 1 minute is:

$$\begin{aligned} N &= N_0\, e^{-kt} \\ &= \left(1.43\times10^{24}\right) e^{-2.94\times10^{-16}\ \text{min}^{-1}\cdot 1\ \text{min}} \\ &= 1.43\times10^{24} \end{aligned}$$

The rate constant is so small that a negligible number of disintegrations occur in one minute.

P36.16) Show that the ratio of the half-life to the three-quarter life, $t_{1/2}/t_{3/4}$ for a reaction that is nth order ($n > 1$) in reactant A can be written as a function of n alone (that is, there is no concentration dependence in the ratio).

In general, for order n,

$$kt = \frac{1}{n-1}\left\{\frac{1}{[A]^{n-1}} - \frac{1}{[A]_0^{n-1}}\right\}$$

$$t_{1/2} \Rightarrow [A] = \frac{1}{2}[A]_0$$

$$kt_{1/2} = \frac{1}{n-1}\left\{\frac{2^{n-1}}{[A]_0^{n-1}} - \frac{1}{[A]_0^{n-1}}\right\}$$

and

$$t_{1/2} = \frac{2^{n-1}-1}{k(n-1)[A]_0^{n-1}}$$

$$t_{3/4} \Rightarrow [A] = \frac{1}{4}[A]_0$$

$$kt_{3/4} = \frac{1}{n-1}\left\{\frac{4^{n-1}}{[A]_0^{n-1}} - \frac{1}{[A]_0^{n-1}}\right\}$$

and

$$t_{3/4} = \frac{4^{n-1}-1}{k(n-1)[A]_0^{n-1}}$$

Thus,

$$\frac{t_{1/2}}{t_{3/4}} = \frac{\dfrac{2^{n-1}-1}{k(n-1)[\mathrm{A}]_0^{\,n-1}}}{\dfrac{4^{n-1}-1}{k(n-1)[\mathrm{A}]_0^{\,n-1}}}$$

$$\frac{t_{1/2}}{t_{3/4}} = \frac{2^{n-1}-1}{4^{n-1}-1}$$

P36.18) For the sequential reaction $\mathrm{A} \xrightarrow{k_A} \mathrm{B} \xrightarrow{k_B} \mathrm{C}$, the rate constants are $k_A = 5 \times 10^6\ \mathrm{s}^{-1}$ and $k_B = 3 \times 10^6\ \mathrm{s}^{-1}$. Determine the time at which [B] is at a maximum.

For a sequential reaction,

$$[\mathrm{B}] = \frac{k_\mathrm{A}}{k_\mathrm{B}-k_\mathrm{A}}\left(e^{-k_\mathrm{A}t} - e^{-k_\mathrm{B}t}\right)[\mathrm{A}_0]$$

The maximum occurs when

$$\frac{\partial[\mathrm{B}]}{\partial t} = 0 = \frac{k_\mathrm{A}}{k_\mathrm{B}-k_\mathrm{A}}[\mathrm{A}]_0 \frac{\partial}{\partial t}\left(e^{-k_\mathrm{A}t} - e^{-k_\mathrm{B}t}\right)$$

$$0 = \frac{k_\mathrm{A}}{k_\mathrm{B}-k_\mathrm{A}}[\mathrm{A}]_0\left(-k_\mathrm{A}e^{-k_\mathrm{A}t} + k_\mathrm{B}e^{-k_\mathrm{B}t}\right)$$

The above equality will be true when the term in parenthesis equals zero, therefore:

$$k_A e^{-k_A t} = k_B e^{-k_B t}$$

$$\ln k_A - k_A t = \ln k_B - k_B t$$

$$\ln k_A - \ln k_B = (k_A - k_B)t$$

$$\frac{1}{k_A - k_B}\ln\left(\frac{k_A}{k_B}\right) = t$$

Substituting the values of k_A and k_B into the previous expression yields:

$$t = \frac{1}{(5\times10^6\ \mathrm{s}^{-1}) - (3\times10^6\ \mathrm{s}^{-1})}\ln\left(\frac{5\times10^6\ \mathrm{s}^{-1}}{3\times10^6\ \mathrm{s}^{-1}}\right)$$

$$= (5\times10^{-7}\ \mathrm{s})(0.511)$$

$$t = 2.55\times10^{-7}\ \mathrm{s}$$

P36.21) For a type II second-order reaction, the reaction is 60% complete in 60 seconds when $[A]_0 = 0.1$ M and $[B]_0 = 0.5$ M.
a) What is the rate constant for this reaction?
b) Will the time for the reaction to reach 60% completion change if the initial reactant concentrations are decreased by a factor of two?

a) The integrated rate-law expression for a second-order reaction of type II is:

$$kt = \frac{1}{[B]_0 - [A]_0} \ln\left(\frac{[B]/[B]_0}{[A]/[A]_0}\right)$$

at t = 60 s, [A] = 0.04 M with 1:1 stoichiometry so that [B] = 0.44 M. Substituting these values into the above expression and using t = 60 s yields:

$$k = \frac{1}{60\text{ s}\,(0.5\text{ M} - 0.1\text{ M})} \ln\left(\frac{0.44\text{ M}/0.5\text{ M}}{0.04\text{ M}/0.1\text{ M}}\right)$$
$$= \left(0.0417\text{ M}^{-1}\text{ s}^{-1}\right)(0.788)$$
$$= 0.0329\text{ M}^{-1}\text{ s}^{-1}$$

b) The time will double, assuming the k value is the same. Numerically checking this expectation:

$$t = \frac{1}{k([B]_0 - [A]_0)} \ln\left(\frac{[B]/[B]_0}{[A]/[A]_0}\right)$$
$$= \frac{1}{\left(0.0329\text{ M}^{-1}\text{ s}^{-1}\right)(0.25\text{ M} - 0.05\text{ M})} \ln\left(\frac{0.22\text{ M}/0.25\text{ M}}{0.02\text{ M}/0.05\text{ M}}\right)$$
$$= (152\text{ s})(0.788)$$
$$= 120\text{ s}$$

P36.24) In the stratosphere, the rate constant for the conversion of ozone to molecular oxygen by atomic chlorine is $Cl + O_3 \rightarrow ClO + O_2$ $[(k = 1.7 \times 10^{10}\text{ M}^{-1}\text{s}^{-1})e^{-260K/T}]$.
a) What is the rate of this reaction at 20 km where $[Cl] = 5 \times 10^{-17}$ M, $[O_3] = 8 \times 10^{-9}$ M, and T = 220 K?

b) The actual concentrations at 45 km are [Cl] = 3×10^{-15} M and [O_3] = 8×10^{-11} M. What is the rate of the reaction at this altitude where T = 270 K?
c) (Optional) Given the concentrations in part (a), what would you expect the concentrations at 45 km to be assuming that the gravity represents the operative force defining the potential energy?

a) Based on the units of k, the reaction is second order overall so that the rate law expression is:

$$Rate = k[\mathrm{Cl}][\mathrm{O_3}]$$

For [Cl] = 5×10^{-17} M, [O_3] = 8×10^{-9} M, and T = 220 K

$$k = 1.7\times10^{10}\ \mathrm{M^{-1}\ s^{-1}}e^{-260\mathrm{K}/220\ \mathrm{K}}$$
$$k = 5.21\times10^{9}\ \mathrm{M^{-1}\ s^{-1}}$$

$$Rate = 5.21\times10^{9}\ \mathrm{M^{-1}\ s^{-1}}\left(5\times10^{-17}\ \mathrm{M}\right)\left(8\times10^{-9}\ \mathrm{M}\right)$$
$$= 2.08\times10^{-15}\ \mathrm{M\ s^{-1}}$$

b) [Cl] = 3×10^{-15} M [O_3] = 8×10^{-11} M T = 270 K

$$k = 1.7\times10^{10}\ \mathrm{M^{-1}\ s^{-1}}e^{-260\mathrm{K}/270\ \mathrm{K}}$$
$$k = 6.49\times10^{9}\ \mathrm{M^{-1}\ s^{-1}}$$

$$Rate = 6.49\times10^{9}\ \mathrm{M^{-1}\ s^{-1}}\left(3\times10^{-15}\ \mathrm{M}\right)\left(8\times10^{-11}\ \mathrm{M}\right)$$
$$= 1.56\times10^{-15}\ \mathrm{M\ s^{-1}}$$

c) The ratio of pressures at two altitudes is given by:

$$\frac{[\ ]_1}{[\ ]_2} = e^{-\frac{mg(h_1-h_2)}{kT}}$$

Using this expression to determine the difference in concentration for Cl at 45 versus 20 km yields:

$$\frac{[Cl]_{45}}{[Cl]_{20}} = e^{-\frac{\left(0.035\ \mathrm{kg\ mol^{-1}}\right)\left(\frac{1}{N_A}\right)\left(9.80\ \mathrm{m\ s^{-2}}\right)\left(2.5\times10^{4}\ \mathrm{km}\right)}{\left(1.38\times10^{-23}\ \mathrm{J\ K^{-1}}\right)\left(270\ \mathrm{K}\right)}} = 0.0219$$

$$[Cl]_{45} = \left(5\times10^{-17}\ \mathrm{M}\right)\left(0.0219\right) = 1.10\times10^{-18}\ \mathrm{M}$$

Performing the same calculation for O_3:

$$\frac{[O_3]_{45}}{[O_3]_{20}} = e^{-\frac{(0.048 \text{ kg mol}^{-1})\left(\frac{1}{N_A}\right)(9.80 \text{ m s}^{-2})(2.5\times10^4 \text{ km})}{(1.38\times10^{-23} \text{ J K}^{-1})(270 \text{ K})}} = 0.0053$$

$$[O_3]_{45} = (8\times10^{-9} \text{ M})(0.0053) = 4.24\times10^{-11} \text{ M}$$

Finally, the rate is:

$$\begin{aligned} Rate &= 6.49\times10^{9} \text{ M}^{-1} \text{ s}^{-1}(1.10\times10^{-18} \text{ M})(4.24\times10^{-11} \text{ M}) \\ &= 3.03\times10^{-19} \text{ M s}^{-1} \end{aligned}$$

Notice that since this simple model for the concentration dependence versus altitude significantly underestimates the concentration of Cl, the rate of ozone depletion by reaction with Cl is also significantly underestimated.

P36.26) A standard "rule of thumb" for thermally activated reactions is that the reaction rate doubles for every 10 K increase in temperature. Is this statement true independent of the activation energy (assuming that the activation energy is positive and independent of temperature)?

The analysis is best performed numerically

$$\frac{k_2}{k_1} = \frac{e^{-E_a/RT_2}}{e^{-E_a/RT_1}} = e^{-\frac{E_a}{R}\left(\frac{1}{T_2}-\frac{1}{T_1}\right)}$$

T_1 (K)	T_2 (K)	E_a (J mol^{-1})	$\frac{k_2}{k_1}$
100	110	50,000	236
300	310	50,000	1.91
1000	1010	50,000	1.06
300	310	500,000	6.43
300	310	5,000	1.06

Inspection of the table demonstrates that the rule is only valid for temperatures around room temperature, and moderate E_a values (~50,000 –70,000 J mol^{-1}).

P36.30) At 552.3 K, the rate constant for the thermal decomposition of SO_2Cl_2 is 1.02×10^{-6} s^{-1}. If the activation energy is 210 kJ mol^{-1}, calculate the Arrhenius preexponential factor and determine the rate constant at 600 K.

$$k_{552.3\ \text{K}} = 1.02\times10^{-6}\ \text{s}^{-1} \qquad \text{E}_a = 210\ \text{kJ mol}^{-1}$$

$$1.02\times10^{-6}\ \text{s}^{-1} = Ae^{-2.10\times10^5\ \text{J mol}^{-1}/\left(8.314\ \text{J mol}^{-1}\ \text{K}^{-1}\right)(552.3\ \text{K})}$$

$$\text{A} = \left(1.02\times10^{-6}\ \text{s}^{-1}\right)e^{2.10\times10^5\ \text{J mol}^{-1}/\left(8.314\ \text{J mol}^{-1}\ \text{K}^{-1}\right)(552.3\ \text{K})}$$

$$\text{A} = 7.38\times10^{13}\ \text{s}^{-1}$$

$$k_{600\ \text{K}} = \left(7.38\times10^{13}\ \text{s}^{-1}\right)e^{-2.10\times10^5\ \text{J mol}^{-1}/\left(8.314\ \text{J mol}^{-1}\text{K}^{-1}\right)(600\ \text{K})}$$

$$= \left(7.38\times10^{13}\ \text{s}^{-1}\right)\left(5.23\times10^{-19}\right)$$

$$= 3.86\times10^{-5}\ \text{s}^{-1}$$

P36.32) Consider the reaction $\text{A} + \text{B} \underset{k'}{\overset{k}{\rightleftharpoons}} \text{P}$ A temperature-jump experiment is performed where the relaxation time constant is measured to be 310 μs, resulting in an equilibrium where $K_{eq} = 0.7$ with $[P]_{eq} = 0.2$ M. What are k and k'? (Watch the units!)

$$\text{A} + \text{B} \underset{k'}{\overset{k}{\rightleftharpoons}} \text{P}$$

$\tau = 310 \times 10^{-6}$ s $\qquad$ $\text{K}_{eg} = 0.7$ $\qquad$ $[\text{P}]_{eg} = 0.2$ M

Assuming the following rate law,

$$\frac{d[\text{A}]}{dt} = -k^{+}[\text{A}][\text{B}] + k^{+'}[\text{P}] = 0,$$

The post-jump equilibrium concentrations with respect to the initial concentrations and concentration shift are:

$[\text{A}] - \xi = [\text{A}]_{eq}$
$[\text{B}] - \xi = [\text{B}]_{eq}$
$[\text{P}] + \xi = [\text{P}]_{eq}$

Therefore, the differential rate expression for the concentration shift, ξ, is:

$$\frac{d\xi}{dt} = -k^{+}\left([\text{A}]_{eq} + \xi\right)\left([\text{B}]_{eq} + \xi\right) + k^{+'}\left([\text{P}]_{eq} - \xi\right)$$

$$\frac{d\xi}{dt} = -\xi\left(k^{+}[\mathrm{A}]_{eq} + k^{+}[\mathrm{B}]_{eq} + k^{+}\xi + k^{+'}\right)$$
$$= -\xi\left(k^{+}\left([\mathrm{A}]_{eq} + k^{+}[\mathrm{B}]_{eq}\right) + k^{+'}\right) + \mathrm{O}\left(\xi^{2}\right)$$

Ignoring terms on the order ξ^2, the relaxation time is:

$$\tau = \left[k^{+}\left([\mathrm{A}]_{eq} + k^{+}[\mathrm{B}]_{eq}\right) + k^{+'}\right]^{-1}$$

Next, the equilibrium constant is given by:

$$K = \frac{k^{+}}{k^{+'}} = \frac{[\mathrm{P}]_{eq}}{[\mathrm{A}]_{eq}[\mathrm{B}]_{eq}} = 0.7$$

If we assume that $[A]_o = [B]_o$ then $[A]_{eg} = [B]_{eg}$ and

$$0.7 = \frac{0.2\ \mathrm{M}}{\mathrm{x}^2} \Rightarrow \mathrm{x} = 0.535\ \mathrm{M} = [\mathrm{A}]_{eq} = [\mathrm{B}]_{eq}$$

And using the expression for K, we know that $k^{+} = 0.7\ k^{+'}$. Use these last two results in the expression for the relaxation time yields:

$$310 \times 10^{-6}\ \mathrm{s} = \frac{1}{k^{+}(0.535 + 0.535) + k^{+'}}$$

$$310 \times 10^{-6}\ \mathrm{s} = \frac{1}{\left(0.7k^{+'}\right)(1.070) + k^{+'}}$$

$$310 \times 10^{-6}\ \mathrm{s} = \frac{1}{k^{+'}(1.749)}$$

$$k^{+'} = 1845\ \mathrm{s}^{-1}$$
$$k^{+} = 0.7k^{+'} = 1291\ \mathrm{M}^{-1}\ \mathrm{s}^{-1}$$

The units of the rate constants are consistent with the forward reaction being second order, and the reverse reaction being first order.

P36.36) Consider the "unimolecular" isomerization of methylcyanide, a reaction that will be discussed in detail in the subsequent chapter:

$$\mathrm{CH_3NC}(g) \longrightarrow \mathrm{CH_3CN}(g)$$

The Arrhenius parameters for this reaction are $A = 2.5 \times 10^{16}\ \text{s}^{-1}$ and $E_a = 272\ \text{kJ mol}^{-1}$. Determine the Eyring parameters $\Delta H^{\ddagger}$ and $\Delta S^{\ddagger}$ for this reaction with $T = 300$ K.

For a unimolecular gas phase reaction

$$E_a = \Delta H^{\ddagger} + RT$$

$$\Delta H^{\ddagger} = 272 \times 10^3\ \text{J mol}^{-1} - \left(8.314\ \text{J mol}^{-1}\ \text{K}^{-1}\right)\left(300\ \text{K}\right)$$

$$= 272 \times 10^3\ \text{J mol}^{-1} - 2.49_4 \times 10^3\ \text{J mol}^{-1}$$

$$= 269.5 \times 10^3\ \text{J mol}^{-1}$$

$$A = \frac{ek_B T}{h} e^{\Delta S^{\ddagger}/R}$$

$$\Delta S^{\ddagger} = R \ln\left(\frac{Ah}{ek_B T}\right)$$

$$= 8.314\ \text{J mol}^{-1}\text{K}^{-1} \ell\text{n}\left(\frac{\left(2.5 \times 10^{16}\ \text{s}^{-1}\right)\left(6.626 \times 10^{-34}\ \text{J s}\right)}{e\left(1.38 \times 10^{-23}\ \text{J K}^{-1}\right)\left(300\ \text{K}\right)}\right)$$

$$= \left(8.314\ \text{J mol}^{-1}\ \text{K}^{-1}\right) \ln\left(1737\right)$$

$$= 62.0\ \text{J mol}^{-1}\text{K}^{-1}$$

P36.38) Chlorine monoxide (ClO) demonstrates three bimolecular self-reactions:

$$Rxn_1:\quad \text{ClO}\bullet(g) + \text{ClO}\bullet(g) \xrightarrow{k_1} \text{Cl}_2(g) + \text{O}_2(g)$$

$$Rxn_2:\quad \text{ClO}\bullet(g) + \text{ClO}\bullet(g) \xrightarrow{k_2} \text{Cl}\bullet(g) + \text{ClOO}\bullet(g)$$

$$Rxn_3:\quad \text{ClO}\bullet(g) + \text{ClO}\bullet(g) \xrightarrow{k_3} \text{Cl}\bullet(g) + \text{OClO}\bullet(g)$$

The following table provides the Arrhenius parameters for this reaction:

	A ($\text{M}^{-1}\ \text{s}^{-1}$)	E_a (kJ/mol)
Rxn_1	6.08×10^8	13.2
Rxn_2	1.79×10^{10}	20.4
Rxn_3	2.11×10^8	11.4

a) For which reaction is $\Delta H^{\ddagger}$ greatest and by how much relative to the next closest reaction?
b) For which reaction is $\Delta S^{\ddagger}$ the smallest and by how much relative to the next closest reaction?

a) These are all bimolecular gas phase reactions; therefore, the relationship between $\Delta H^{\ddagger}$ and E_a is equivalent. Since each $\Delta H^{\ddagger}$ depends linearly on the T and E_a, at the same T, the largest E_a corresponds to the largest $\Delta H^{\ddagger}$. Therefore, Rxn_2 will have the largest $\Delta H^{\ddagger}$ by 7.2 kJ mol^{-1} relative to Rxn_1.

b) $\Delta S^{\ddagger}$ depends linearly on $\ln A$; therefore, the smallest $\Delta S^{\ddagger}$ corresponds to the reaction with the smallest A, or Rxn_3. The next smallest $\Delta S^{\ddagger}$ is Rxn_1, and the difference is:

$$\ln\left(6.08\times10^{8}\right)-\ln\left(2.11\times10^{8}\right)=\ln\left(\frac{6.08}{2.11}\right)=\ln\left(2.88\right)=1.06$$

Chapter 37: Complex Reaction Mechanisms

P37.1) A proposed mechanism for the formation of N_2O_5 from NO_2 and O_3 is

$$NO_2 + O_3 \xrightarrow{k_1} NO_3 + O_2$$

$$NO_3 + NO_2 + M \xrightarrow{k_2} N_2O_5 + M$$

Determine the rate law expression for the production of N_2O_5 given this mechanism.

$$\frac{d[N_2O_5]}{dt} = k_2[NO_2][NO_3]$$

$$\frac{d[NO_3]}{dt} = k_2[NO_2][O_3] - k_2[NO_2][NO_3]$$

Applying the steady state approximation to the intermediate NO_3 and substituting back into the differential rate expression for N_2O_5 yields:

$$\frac{d[NO_3]}{dt} = 0 = k_1[NO_2][O_3] - k_2[NO_2][NO_3]$$

$$k_2[NO_2][NO_3] = k_1[NO_2][O_3]$$

$$[NO_3] = \frac{k_1}{k_2}[O_3]$$

$$\frac{d[N_2O_5]}{dt} = k_2[NO_2][NO_3]$$

$$= k_2[NO_2]\left(\frac{k_1}{k_2}[O_3]\right)$$

$$= k_1[NO_2][O_3]$$

The mechanism predicts that the reaction is first order in NO_2 and O_3, second order overall.

P37.4) The hydrogen-bromine reaction corresponds to the production of HBr from H_2 and Br_2 as follows: $H_2 + Br_2 \rightleftarrows 2HBr$. This reaction is famous for its complex rate law, determined by Bodenstein and Lind in 1906:

$$\frac{d[HBr]}{dt} = \frac{k[H_2][Br_2]^{1/2}}{1 + \frac{m[HBr]}{[Br_2]}}$$

where k and m are constants. It took 13 years for the correct mechanism of this reaction to be proposed, and this feat was accomplished simultaneously by Christiansen, Herzfeld, and Polyani. The mechanism is as follows:

$$Br_2 \underset{k_{-1}}{\overset{k_1}{\rightleftharpoons}} 2Br\cdot$$

$$Br\cdot + H_2 \xrightarrow{k_2} HBr + H\cdot$$
$$H\cdot + Br_2 \xrightarrow{k_3} HBr + Br\cdot$$
$$HBr + H\cdot \xrightarrow{k_4} H_2 + Br\cdot$$

Construct the rate law expression for the hydrogen-bromine reaction by performing the following steps:

a) Write down the differential rate expression for [HBr].
b) Write down the differential rate expressions for [Br] and [H].
c) Because Br and H are reaction intermediates, apply the steady-state approximation to the result of part (b).
d) Add the two equations from part (c) to determine [Br] in terms of [Br_2].
e) Substitute the expression for [Br] back into the equation for [H] derived in part (c) and solve for [H].
f) Substitute the expressions for [Br] and [H] determined in part (e) into the differential rate expression for [HBr] to derive the rate law expression for the reaction.

a) $$\frac{d[HBr]}{dt} = k_2[Br][H_2] + k_3[H][Br_2] - k_4[HBr][H]$$

b) $$\frac{d[Br]}{dt} = 2k_1[Br_2] - 2k_{-1}[Br]^2 - k_2[Br][H_2] + k_3[H][Br_2] + k_4[HBr][H]$$
$$\frac{d[H]}{dt} = k_2[Br][H_2] - k_3[H][Br_2] - k_4[HBr][H]$$

c) Applying the steady state approximation to [Br•] and [H•] and adding yields

$$0 = 2k_1[Br_2] - 2k_{-1}[Br]^2 - k_2[Br][H_2] + k_3[H][Br_2] + k_4[HBr][H]$$
$$0 = k_2[Br][H_2] - k_3[H][Br_2] - k_4[HBr][H]$$

d)
$$0 = 2k_1[Br_2] - 2k_{-1}[Br]^2$$
$$[Br] = \sqrt{\frac{k_1}{k_{-1}}}[Br_2]^{1/2}$$

e)

$$k_2[\text{Br}][\text{H}_2]=k_3[\text{H}][\text{Br}_2]+k_4[\text{HBr}][\text{H}]$$

$$\frac{k_2[\text{Br}][\text{H}_2]}{k_3[\text{Br}_2]+k_4[\text{HBr}]}=[\text{H}]$$

$$\frac{k_2\sqrt{\frac{k_1}{k_{-1}}}[\text{Br}_2]^{1/2}[\text{H}_2]}{k_3[\text{Br}_2]+k_4[\text{HBr}]}=[\text{H}]$$

f) Now, substitution into (a) yields

$$\frac{d[\text{HBr}]}{dt}=k_2[\text{Br}][\text{H}_2]+k_3[\text{H}][\text{Br}_2]-k_4[\text{HBr}][\text{H}]$$

$$=k_2\sqrt{\frac{k_1}{k_{-1}}}[\text{Br}_2]^{1/2}[\text{H}_2]+\frac{k_3k_2\sqrt{\frac{k_1}{k_{-1}}}[\text{Br}_2]^{3/2}[\text{H}_2]}{k_3[\text{Br}_2]+k_4[\text{HBr}]}-\frac{k_4k_2\sqrt{\frac{k_1}{k_{-1}}}[\text{Br}_2]^{1/2}[\text{H}_2][\text{HBr}]}{k_3[\text{Br}_2]+k_4[\text{HBr}]}$$

$$=k_2\sqrt{\frac{k_1}{k_{-1}}}[\text{Br}_2]^{1/2}[\text{H}_2]\left(\frac{k_3[\text{Br}_2]+k_4[\text{HBr}]}{k_3[\text{Br}_2]+k_4[\text{HBr}]}\right)+\frac{k_3k_2\sqrt{\frac{k_1}{k_{-1}}}[\text{Br}_2]^{3/2}[\text{H}_2]}{k_3[\text{Br}_2]+k_4[\text{HBr}]}$$

$$-\frac{k_4k_2\sqrt{\frac{k_1}{k_{-1}}}[\text{Br}_2]^{1/2}[\text{H}_2][\text{HBr}]}{k_3[\text{Br}_2]+k_4[\text{HBr}]}$$

$$=\frac{2k_3k_2\sqrt{\frac{k_1}{k_{-1}}}[\text{Br}_2]^{3/2}[\text{H}_2]}{k_3[\text{Br}_2]+k_4[\text{HBr}]}$$

$$=\frac{2k_2\sqrt{\frac{k_1}{k_{-1}}}[\text{Br}_2]^{1/2}[\text{H}_2]}{1+\frac{k_4[\text{HBr}]}{k_3[\text{Br}_2]}}$$

P37.6) For the reaction $\text{I}^-(aq)+\text{OCl}^-(aq)\rightleftarrows\text{OI}^-(aq)+\text{Cl}^-(aq)$ occurring in aqueous solution, the following mechanism has been proposed:

$$\text{OCl}^-+\text{H}_2\text{O}\underset{k_{-1}}{\overset{k_1}{\rightleftarrows}}\text{HOCl}+\text{OH}^-$$

$$\text{I}^-+\text{HOCl}\xrightarrow{k_2}\text{HOI}+\text{Cl}^-$$

$$\text{HOI}+\text{OH}^-\xrightarrow{k_3}\text{H}_2\text{O}+\text{OI}^-$$

a) Derive the rate law expression for this reaction based on this mechanism. (*Hint:* $[OH^-]$ should appear in the rate law.)

b) The initial rate of reaction was studied as a function of concentration by Chia and Connick [*J. Phys. Chem.* 63 (1959), 1518], and the following data were obtained:

$[I^-]_0$ (M)	$[OCl^-]_0$ (M)	$[OH^-]_0$ (M)	Initial Rate (M s^{-1})
2.0×10^{-3}	1.5×10^{-3}	1.00	1.8×10^{-4}
4.0×10^{-3}	1.5×10^{-3}	1.00	3.6×10^{-4}
2.0×10^{-3}	3.0×10^{-3}	2.00	1.8×10^{-4}
4.0×10^{-3}	3.0×10^{-3}	1.00	7.2×10^{-4}

Is the predicted rate law expression derived from the mechanism consistent with these data?

a) $$\frac{d\left[\mathrm{OI}^-\right]}{dt} = k_3[\mathrm{HOI}]\left[\mathrm{OH}^-\right]$$

The intermediate species have rate expressions (with steady state approximation)

$$\frac{d[\mathrm{HOCl}]}{dt} = 0 = k_1\left[\mathrm{OCl}^-\right][\mathrm{H_2O}] - k_1[\mathrm{HOCl}]\left[\mathrm{OH}^-\right] - k_2\left[\mathrm{I}^-\right][\mathrm{HOCl}]$$

$$\frac{d[\mathrm{HOI}]}{dt} = k_2\left[\mathrm{I}^-\right][\mathrm{HOCl}] - k_3[\mathrm{HOI}]\left[\mathrm{OH}^-\right] = 0$$

Solving the last expression for [HOI] and substituting into the differential rate expression for $[OI^-]$ yields:

$$\frac{d\left[\mathrm{OI}^-\right]}{dt} = k_2\left[\mathrm{I}^-\right][\mathrm{HOCl}]$$

Next, [HOCl] must be expressed in terms of reactants and $[OH^-]$ Rearranging the steady-state approximation applied to the differential rate expression for [HOCl] results in the following:

$$[\mathrm{HOCl}] = \frac{k_1\left[\mathrm{OCl}^-\right][\mathrm{H_2O}]}{k_1\left[\mathrm{OH}^-\right] + k_2\left[\mathrm{I}^-\right]}$$

This result is substituted into the differential rate expression for $[OI^-]$ to yield:

$$\frac{d\left[\mathrm{OI}^-\right]}{dt} = \frac{k_1 k_2\left[\mathrm{OCl}^-\right][\mathrm{H_2O}]\left[\mathrm{I}^-\right]}{k_1\left[\mathrm{OH}^-\right] + k_2\left[\mathrm{I}^-\right]}$$

Inspection of the concentrations employed in the table suggests that $k_2[I^-] << k_1[OI^-]$ resulting in:

$$\frac{d\left[OI^-\right]}{dt} = \frac{k_2\left[I^-\right]\left[OCI^-\right]\left[H_2O\right]}{\left[OH^-\right]}$$

A markedly similar expression is obtained using the pre-equilibrium approximation to determine[HOCl]

b) Consider set 1 & 2: $[I^-]_o$ is doubled, doubling the rate.
set 1 & 3: Doubling $[OCI^-]$ and $[OH^-]$ results in no net change in rate.
set 1 & 4: Doubling $[I^-]$ and $[OCI^-]$ quadruples the rate.

These results confirm the rate expression.

P37.8) Consider the following mechanism, which results in the formation of product P:

$$A \underset{k_{-1}}{\overset{k_1}{\rightleftarrows}} B \underset{k_{-2}}{\overset{k_2}{\rightleftarrows}} C$$

$$B \xrightarrow{k_3} P$$

If only the species A is present at $t = 0$, what is the expression for the concentration of P as a function of time? You can apply the preequilibrium approximation in deriving your answer.

Using the pre-equilibrium approximation, we can express [B] and [C] in terms of [A] as follows:

$$[B] = \frac{k_1}{k_{-1}}[A] = K_1[A]$$

$$[C] = \frac{k_2}{k_{-2}}[B] = K_2[B] = K_1K_2[A]$$

$$\frac{d[P]}{dt} = k_3[B] = k_3K_1[A]$$

Next, from mass conservation:

$$[\mathrm{A}]_0 = [\mathrm{A}]+[\mathrm{B}]+[\mathrm{C}]+[\mathrm{P}]$$

$$\frac{d[\mathrm{A}]_0}{dt} = 0 = \frac{d[\mathrm{A}]}{dt}+\frac{d[\mathrm{B}]}{dt}+\frac{d[\mathrm{C}]}{dt}+\frac{d[\mathrm{P}]}{dt}$$

$$\frac{d[\mathrm{P}]}{dt} = -\left(\frac{d[\mathrm{A}]}{dt}+\frac{d[\mathrm{B}]}{dt}+\frac{d[\mathrm{C}]}{dt}\right)$$

$$\frac{d[\mathrm{P}]}{dt} = -\left(\frac{d[\mathrm{A}]}{dt}+\frac{d(K_1[\mathrm{A}])}{dt}+\frac{d(K_1K_2[\mathrm{A}])}{dt}\right)$$

$$= -(1+K_1+K_1K_2)\frac{d[\mathrm{A}]}{dt}$$

Setting the two differential rate expressions for [P] and integrating yields:

$$-(1+K_1+K_1K_2)\frac{d[\mathrm{A}]}{dt} = k_3K_1[\mathrm{A}]$$

$$\int_{[\mathrm{A}]_0}^{[\mathrm{A}]}\frac{d[\mathrm{A}]}{[\mathrm{A}]} = \int_0^t \frac{-k_3K_1}{(1+K_1+K_1K_2)}dt$$

$$[\mathrm{A}] = [\mathrm{A}]_0 e^{\frac{-k_3K_1t}{(1+K_1+K_1K_2)}}$$

If a pre-equilibrium is established rapidly before any product formation:

$$[\mathrm{A}]_0 = [\mathrm{A}]_0^{eq}+[\mathrm{B}]_0^{eq}+[\mathrm{C}]_0^{eq}$$

$$= (1+K_1+K_1K_2)[\mathrm{A}]_0^{eq}$$

$$[\mathrm{P}] = [\mathrm{A}]_0 - ([\mathrm{A}]+[\mathrm{B}]+[\mathrm{C}])$$

$$= [A]_0 - ([\mathrm{A}]+K_1[\mathrm{A}]+K_1K_2[\mathrm{A}])$$

$$= [A]_0 - (1+K_1+K_1K_2)[\mathrm{A}]$$

$$= [A]_0 - (1+K_1+K_1K_2)\frac{[\mathrm{A}]_0}{(1+K_1+K_1K_2)}e^{\frac{-k_3K_1t}{(1+K_1+K_1K_2)}}$$

$$= [A]_0\left(1-e^{\frac{-k_3K_1t}{(1+K_1+K_1K_2)}}\right)$$

P37.12) The enzyme fumarase catalyzes the hydrolysis of fumarate:

$$\text{Fumarate} + H_2O \longrightarrow \text{L-malate}$$

The turnover number for this enzyme is $2.5 \times 10^3\ s^{-1}$, and the Michaelis constant is 4.2×10^{-6} M. What is the rate of fumarate conversion if the initial enzyme concentration is 1×10^{-6} M and the fumarate concentration is 2×10^{-4} M?

Recognizing that k_2 is the turnover number, the values provided in the problem can be used directly in the rate expression to determine the initial rate of reaction:

$k_2 = 2.5 \times 10^3\ s_{-1}$ $\qquad K_m = 4.2 \times 10^{-6}$ M
$[E]_0 = 1 \times 10^{-6}$ M $\qquad [S]_0 = 2 \times 10^{-4}$ M

$$rate_0 = \frac{k_2[\mathrm{S}]_0[\mathrm{E}]_0}{[\mathrm{S}]_0 + K_m}$$

$$= \frac{(2.5\times10^3\ \mathrm{s}^{-1})(2\times10^{-4}\ \mathrm{M})(1\times10^{-6}\ \mathrm{M})}{(2\times10^{-4}\ \mathrm{M})+(4.2\times10^{-6}\ \mathrm{M})}$$

$$= 2.45\times10^{-3}\ \mathrm{M\ s^{-1}}$$

P37.16) The enzyme glycogen synthase kinase (GSK-3β) plays a central role in Alzheimer's disease. The onset of Alzheimer's disease is accompanied by the production of highly phosphorylated forms of a protein referred to as "τ." GSK-3β contributes to the hyperphosphorylation of τ such that inhibiting the activity of this enzyme represents a pathway for the development of an Alzheimer's drug. A compound known as Ro 31-8220 is a competitive inhibitor of GSK-3β. The following data were obtained for the rate of GSK-3β activity in the presence and absence of Ro 31-8220 [A. Martinez *et al., J. Medicinal Chemistry* 45 (2002), 1292]:

$[S]$ (μM)	$Rate_0$ (μM s^{-1}), $[I] = 0$	$Rate_0$ (μM s^{-1}) $[I] = 200\ \mu$M
66.7	4.17×10^{-8}	3.33×10^{-8}
40.0	3.97×10^{-8}	2.98×10^{-8}
20.0	3.62×10^{-8}	2.38×10^{-8}
13.3	3.27×10^{-8}	1.81×10^{-8}
10.0	2.98×10^{-8}	1.39×10^{-8}
6.67	2.31×10^{-8}	1.04×10^{-8}

Determine K_m and $rate_{max}$ for GSK-3β and, using the data with the inhibitor, determine K_m* and K_i.

Analyzing the data without inhibitor using a Lineweaver-Burk plot yields:

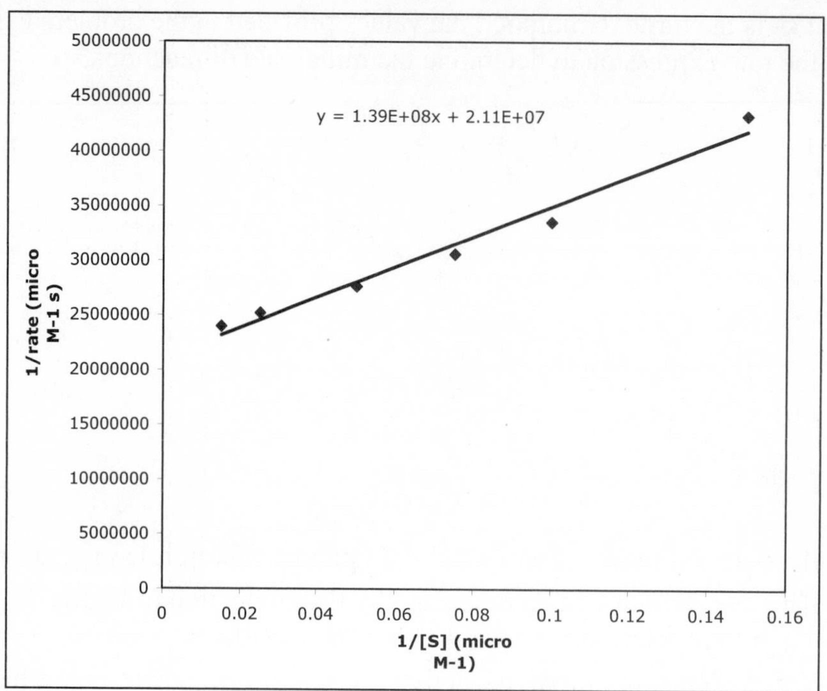

The best-fit straight line to the data yields the following equation:

$$\frac{1}{\text{rate}_0} = 1.39\times10^{8}\ \text{s}\ \frac{1}{[\text{S}]_0} + 2.11\times10^{6}\,\mu\text{M}^{-1}\ \text{s}$$

The maximum rate is equal to the inverse of the y-intercept:

$$\text{rate}_{\text{max}} = \frac{1}{y-\text{int}} = \frac{1}{2.11\times10^{7}\,\mu\text{M}^{-1}\text{s}}$$

$$\text{rate}_{\text{max}} = 4.74\times10^{-8}\ \mu\text{M s}^{-1}$$

With the maximum rate and slope of the best-fit line, the Michaelis constant obtained as follows:

$$K_m = (\text{slope})(\ \text{rate}_{\text{max}}) = \left(1.39\times10^{8}\ \text{s}\ \right)\left(4.74\times10^{-8}\ \mu\text{M s}^{-1}\right)$$

$$K_m = 6.49\ \mu\text{M}$$

Using the inhibited data, the Lineweaver-Burk plot is:

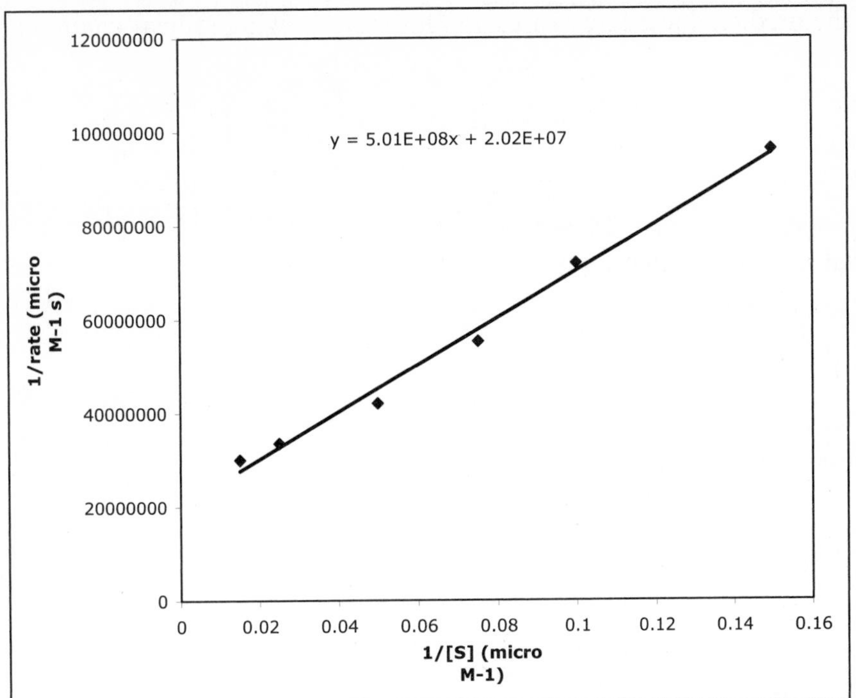

Best fit by a straight line to the data yields the following equation:

$$\frac{1}{\text{rate}_0} = 5.01\times10^8 \text{ s} \frac{1}{[\text{S}]_0} + 2.01\times10^7 \,\mu\text{M}^{-1}\text{s}$$

The maximum rate with inhibitor is equal to the inverse of the y-intercept:

$$\text{rate}_{\text{max}} = \frac{1}{2.01\times10^7 \,\mu\text{M}^{-1}\text{ s}}$$
$$= 4.98\times10^{-8} \,\mu\text{M}^{-1}\text{ s}$$

The apparent Michaelis constant is given by:

$$K_m^* = (\text{slope})(\text{rate}_{\text{max}})$$
$$= \left(5.01\times10^8 \text{ s}\right)\left(4.98\times10^{-8} \,\mu\text{M}^{-1}\text{ s}\right)$$
$$K_m^* = 24.9 \,\mu\text{M}$$

Finally, the K_I value is given by:

$$K_I = \frac{[I]}{\frac{K_m^*}{K_m} - 1} = \frac{200\mu\text{M}}{\left(\frac{24.9 \,\mu\text{M}}{6.49 \,\mu\text{M}}\right) - 1}$$
$$= 70.4 \,\mu\text{M}$$

P37.18) Determine the predicted rate law expression for the following radical-chain reaction:

$$A_2 \xrightarrow{k_1} 2A\cdot$$
$$A\cdot \xrightarrow{k_2} B\cdot + C$$
$$A\cdot + B\cdot \xrightarrow{k_3} P$$
$$A\cdot + P \xrightarrow{k_4} B\cdot$$

The differential rate of P formation is

$$\frac{d[\mathrm{P}]}{dt} = k_3[\mathrm{A}\cdot][\mathrm{B}\cdot] - k_4[\mathrm{A}\cdot][\mathrm{P}]$$

The rate expression for A· and B· are

$$\frac{d[\mathrm{A}\cdot]}{dt} = 2k_1[\mathrm{A}_2] - k_2[\mathrm{A}\cdot] - k_3[\mathrm{A}\cdot][\mathrm{B}\cdot] - k_4[\mathrm{A}\cdot][\mathrm{P}]$$

$$\frac{d[\mathrm{B}\cdot]}{dt} = k_2[\mathrm{A}\cdot] - k_3[\mathrm{A}\cdot][\mathrm{B}\cdot] + k_4[\mathrm{A}\cdot][\mathrm{P}]$$

Applying the steady state approximation for [B]

$$k_3[\mathrm{A}\cdot][\mathrm{B}\cdot] - k_4[\mathrm{A}\cdot][\mathrm{P}] = k_2[\mathrm{A}\cdot]$$
$$[\mathrm{B}\cdot] = \frac{k_2 + k_4[\mathrm{P}]}{k_3}$$

Substituting this result into the differential rate expression for P yields:

$$\frac{d[\mathrm{P}]}{dt} = k_3[\mathrm{A}\cdot]\left(\frac{k_2 + k_4[\mathrm{P}]}{k_3}\right) - k_4[\mathrm{A}\cdot][\mathrm{P}]$$
$$= k_2[\mathrm{A}\cdot]$$

Now, adding the steady-state expressions for $[\mathrm{A}\cdot]$ and $[\mathrm{B}\cdot]$ yields:

$$0 = 2k_2[\mathrm{A}_2] - 2k_3[\mathrm{A}\cdot][\mathrm{B}\cdot]$$
$$[\mathrm{A}\cdot] = \frac{k_2[\mathrm{A}_2]}{k_3[\mathrm{B}\cdot]} = \frac{k_2[\mathrm{A}_2]}{k_2 + k_4[\mathrm{P}]}$$

Substituting this expression into the differential rate expression for P yields the final result:

$$\frac{d[\mathrm{P}]}{dt}=\frac{k_2 k_1[\mathrm{A}_2]}{k_2+k_4[\mathrm{P}]}$$

P37.22) The adsorption of ethyl chloride on a sample of charcoal at 0°C measured at several different pressures is as follows:

$P_{C_2H_5Cl}$ (Torr)	V_{ads} (mL)
20	3.0
50	3.8
100	4.3
200	4.7
300	4.8

Using the Langmuir isotherm, determine the fractional coverage at each pressure and V_M.

The Langmuir equation can be written as

$$\frac{P}{V}=\frac{P}{V_m}+\frac{1}{KV_m}$$

And the fractional coverage, θ, is simply the ratio of adsorbed volume to the volume of maximum adsorption (V_m):

$$\theta=\frac{V}{V_m}$$

Therefore, V_m is required to determine the fractional coverage versus pressure. The plot of $\frac{\mathrm{P}}{\mathrm{V}}$ vs. P should yield a straight line with slope equal to the inverse of V_m. This plot is as follows:

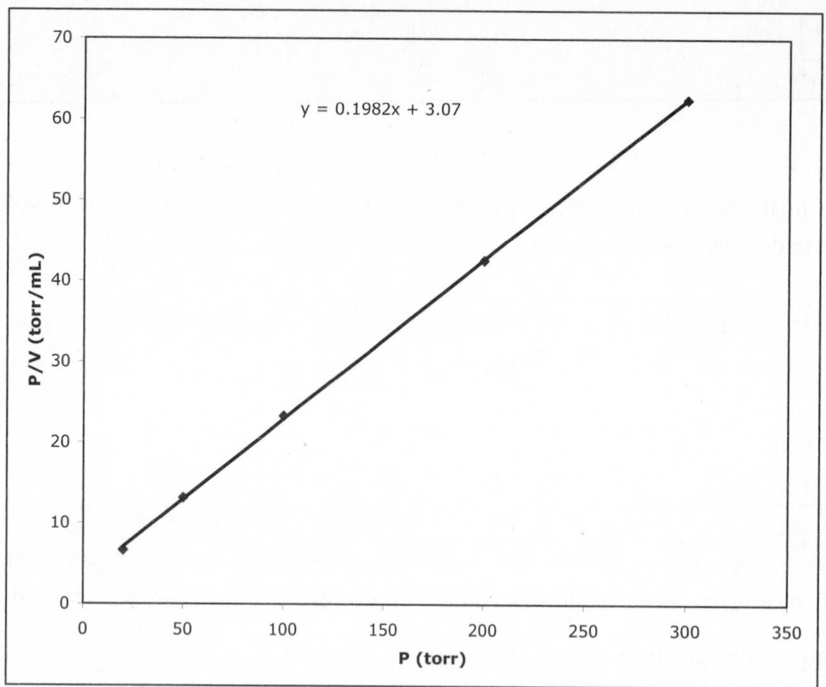

The equation for the best-fit line is: $\frac{P}{V} = 0.198 \text{ mL}^{-1}(P) + 3.07 \text{ torr mL}^{-1}$

Thus, the V_m value is

$$V_m = \frac{1}{slope}$$

$$V_m = 5.04 \text{ mL}$$

With V_m θ can be determined resulting in the following:

P (atm)	θ
20	0.595
50	0.754
100	0.853
200	0.932
300	0.952

P37.25) Many surface reactions require the adsorption of two or more different gases. For the case of two gases, assuming that the adsorption of a gas simply limits the number of surface sites available for adsorption, and derive expressions for the fractional coverage of each gas.

If we assume that the only effect of the gases is to limit the number of sites available for adsorption, then the Langmuir model yields the two following relationships:

$$k_{d1}\theta_1 = k_{a1}P_1(1-\theta_1-\theta_2)$$
$$k_{d2}\theta_2 = k_{a2}P_2(1-\theta_1-\theta_2)$$

Where k_a, k_d, and θ are the rate of adsorption, desorption, and fractional coverage, respectively. In addition, the subscript denotes either gas 1 or 2 in the two gas mixture. The two equations can be solved to yield θ_1 and θ_2 as follows:

$$\theta_1 = \frac{K_1P_1}{1+K_1P_1+K_2P_2}$$

$$\theta_2 = \frac{K_2P_2}{1+K_1P_1+K_2P_2}$$

P37.28) If $\tau_f = 1\times 10^{-10}$ s and $k_{ic} = 5\times 10^{8}$ s^{-1}, what is ϕ_f? Assume that the rate constants for intersystem crossing and quenching are sufficiently small that these processes can be neglected.

Φ_f is related to k_f and τ_f by the expression

$$\Phi_f = k_f\tau_f$$

and k_f is related to τ_f by

$$\frac{1}{\tau_f} = k_f + k_{ic} \qquad \text{(assuming } k_{isc}^3, k_7 \simeq 0\text{)}$$

$$\tau_f = 1\times 10^{-10}\text{ s and } k_{ic} = 5\times 10^{8}\text{ s}^{-1}$$

Thus

$$\frac{1}{1\times 10^{-10}\text{ s}} = k_f + 5\times 10^{8}\text{ s}^{-1}$$

and

$$k_f = 1\times 10^{10}\text{ s}^{-1} - 5\times 10^{8}\text{ s}^{-1}$$
$$k_f = 9.5\times 10^{9}\text{ s}^{-1}$$

And finally,

$$\Phi_f = k_f\tau_f$$
$$= 9.5\times 10^{9}\text{ s}^{-1}\cdot 1\times 10^{-10}$$
$$\Phi_f = 0.95$$

P37.30) If 10% of the energy of a 100-W incandescent bulb is in the form of visible light having an average wavelength of 600 nm, how many quanta of light are emitted per second from the light bulb?

Assuming that 10% of energy from the bulb is in the form of visible light, the power of the light is given by:

$$\begin{aligned} P_\ell &= 0.1(P) \\ &= 0.1(100\text{ W}) \\ &= 10\text{ J s}^{-1} \end{aligned}$$

The energy per photon at 600 nm is:

$$E = \frac{hc}{\lambda} = \frac{(6.626\times10^{-34}\text{ J s})(3.00\times10^{8}\text{ m s}^{-1})}{600\times10^{-9}\text{ m}} = 3.31\times10^{-19}\text{ J}$$

Therefore, the number of photons per second is given by:

$$n = \frac{10\text{ J s}^{-1}}{3.13\times10^{-19}\text{ J photon}^{-1}} = 3.02\times10^{19}\text{ photon s}^{-1}$$

P37.33) A central issue in the design of aircraft is improving the lift of aircraft wings. To assist in the design of more efficient wings, wind-tunnel tests are performed in which the pressures at various parts of the wing are measured generally using only a few localized pressure sensors. Recently, pressure-sensitive paints have been developed to provide a more detailed view of wing pressure. In these paints, a luminescent molecule is dispersed into a oxygen-permeable paint and the aircraft wing is painted. The wing is placed into an airfoil, and luminescence from the paint is measured. The variation in O_2 pressure is measured by monitoring the luminescence intensity, with lower intensity demonstrating areas of higher O_2 pressure due to quenching.

a) The use of platinum octaethylprophyrin (PtOEP) as an oxygen sensor in pressure-sensitive paints was described by Gouterman and coworkers [*Review of Scientific Instruments* 61 (1990), 3340]. In this work, the following relationship between luminescence intensity and pressure was derived: $\frac{I_0}{I} = A + B\left(\frac{P}{P_0}\right)$, where I_0 is the fluorescence intensity at ambient pressure P_0, and I is the fluorescence intensity at an arbitrary pressure P. Determine coefficients A and B in the preceding expression using the Stern–Volmer equation: $k_{total} = \frac{1}{\tau_l} = k_l + k_q[Q]$. In this equation τ_l is the luminescence lifetime, k_r is the luminescent rate constant, and k_q is the quenching rate constant. In addition, the luminescent intensity ratio is equal to the ratio of luminescence quantum yields at ambient pressure, Φ_0, and an arbitrary pressure, Φ: $\frac{\Phi_0}{\Phi} = \frac{I_0}{I}$.

b) Using the following calibration data of the intensity ratio versus pressure observed for PtOEP, determine A and B:

I_0/I	P/P_0	I_0/I	P/P_0
1.0	1.0	0.65	0.46
0.9	0.86	0.61	0.40
0.87	0.80	0.55	0.34
0.83	0.75	0.50	0.28
0.77	0.65	0.46	0.20
0.70	0.53	0.35	0.10

c) At an ambient pressure of 1 atm, $I_0 = 50,000$ (arbitrary units) and 40,000 at the front and back of the wing. The wind tunnel is turned on to a speed of Mach 0.36 and the measured luminescence intensity is 65,000 and 45,000 at the respective locations. What is the pressure differential between the front and back of the wing?

a) Starting with the version of the Stern-Volmer Eq. provided in the problem:

$$k_{tot} = k_l + k_q[\mathrm{Q}]$$

The luminescence quantum yield can be expressed in terms of k_l and k_{total} as:

$$\Phi = \frac{k_l}{k_{total}}$$

Therefore:

$$\frac{\Phi_0}{\Phi} = \frac{k_{total}}{k_{total_0}}$$

$$= \frac{k_l + k_q P}{k_l + k_q P_0}$$

$$= \frac{k_l}{k_l + k_q P_0} + \frac{k_q P_0}{k_l + k_q P_0}\left(\frac{P}{P_0}\right)$$

$$= A + B\left(\frac{P}{P_0}\right)$$

b) The plot of (I_0/I) versus (P/P_0) is as follows:

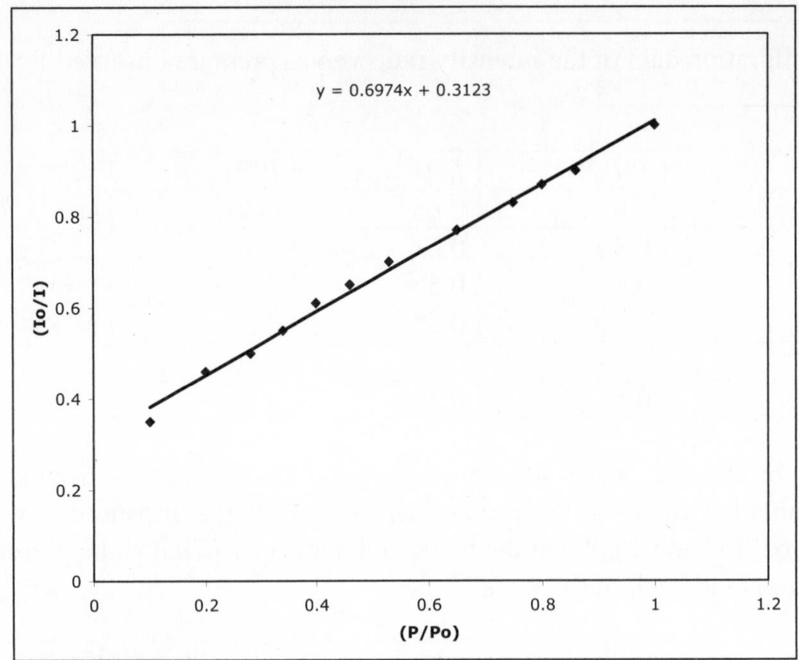

Best fit by a straight line to the data yields the following equation:

$$\frac{I_o}{I} = 0.697\left(\frac{P}{P_0}\right) + 0.312$$

Thus, A = 0.312 and B =0.697.

c)

$$\left(\left(\frac{I_0}{I}\right)_{front} - \left(\frac{I_0}{I}\right)_{back} = \right) = 0.697\left(\frac{P_{front} - P_{back}}{P_0}\right)$$

$$\left(\frac{50,000}{65,000} - \frac{40,000}{45,000}\right) = 0.697\left(\frac{P_{front} - P_{back}}{1 \text{ atm}}\right)$$

$$-0.120 = 0.697\left(\frac{P_{front} - P_{back}}{1 \text{ atm}}\right)$$

$$-0.172 \text{ atm} = P_{front} - P_{back}$$